501
Measurement and Conversion
Questions

501

Measurement and Conversion Questions

LearningExpress ®

NEW YORK

Library of Congress Cataloging-in-Publication Data:
501 measurement and conversion questions / LearningExpress.— 1st ed.
 p. cm.
 ISBN 1-57685-520-1
 1. Measurement—Problems, exercises, etc. I. Title: Five hundred and one
measurement and conversion questions. II. Title: Five hundred one measurement
and conversion. III. LearningExpress (Organization)
 QA465.A14 2004
 516'.15—dc22

                                                        2004003125

Printed in the United States of America

9 8 7 6 5 4 3 2 1

First Edition

ISBN 1-57685-520-1

For more information or to place an order, contact LearningExpress at:
    55 Broadway
    8th Floor
    New York, NY 10006

Or visit us at:
    www.learnatest.com

The LearningExpress Skill Builder in Focus Writing Team is comprised of experts in test preparation, as well as educators and teachers who specialize in language arts and math.

LearningExpress Skill Builder in Focus Writing Team

Marco A. Annunziata
Freelance Writer
New York, New York

Lara Bohlke
Middle School Math Teacher, Grade 8
Dodd Middle School
Cheshire, Connecticut

Elizabeth Chesla
English Instructor
Coordinator of Technical and Professional Communication Program
Polytechnic University, Brooklyn
South Orange, New Jersey

Ashley Clarke
Former Math and Science Teacher
Math Consultant
San Diego, California

Brigit Dermott
Freelance Writer
English Tutor, New York Cares
New York, New York

Darren Dunn
English Teacher
Riverhead School District
Riverhead, New York

Barbara Fine
English Instructor
Secondary Reading Specialist
Setauket, New York

Sandy Gade
Project Editor
LearningExpress
New York, New York

Contents

Introduction

Why Should I Use This Book?

This book was created to provide you with targeted review and practice in the most essential measurement and conversions skills. It provides 501 problems to help solidify your understanding of a variety of concepts. *501 Measurement and Conversion Questions* is designed for many audiences, because measurement and conversion questions are not just found in the classroom or on standardized tests, they abound in every aspect of life and in virtually every profession. If you plan on entering one of the trades, you know how critical good measurement and conversion skills are, but did you know that these skills are just as essential for being a firefighter, a businessperson, a cashier, or a nurse? And the list goes on and on. In addition, having a solid grasp of the knowledge contained in this book is fundamental to scoring higher on the math portions of the SAT, ACT, GRE, or on a vocational or professional exam. And of course, *501 Measurement and Conversion Questions* is great for anyone who has ever taken a math course and wants to refresh and revive forgotten skills. Or, it can be used to supplement current instruction in a math class.

How to Use This Book

First, look at the table of contents to see the types of measurement and conversion topics covered in this book. The book is organized into four measurement chapters and six conversion chapters. The measurements chapters are Perimeter, Area, Volume, and Angles and Arc. The conversion chapters are Dollars and Cents; Fractions, Decimals, and Percents; Scientific Notation and Decimals; Metric and Standard Units: Length, Area, and Volume; and Metric and Standard Units: Temperature and Weight. The structure of these chapters follows a common sequence of math concepts. You may want to follow the sequence because the concepts grow more advanced as the book progresses. However, if your skills are just rusty, or if you are using this book to supplement topics you are currently learning, you may want to jump around from topic to topic.

As you complete the problems in this book, you will undoubtedly want to check your answers against the answer explanation section at the end of each chapter. Every problem in *501 Measurement and Conversion Questions* has a complete answer explanation. For problems that require numerous steps, a thorough step-by-step explanation is provided. This will help you understand the problem-solving process. The purpose of drill and skill practice is to make you proficient at solving problems. Like an athlete preparing for the next season or a musician warming up for a concert, you become more skillful with practice. If, after completing all the problems in a section, you feel you need more practice, do the problems over. It's not the answer that matters most—it's the process and the reasoning skills that you want to master.

You will probably want to have a calculator handy as you work through some of the sections. It's always a good idea to use it to check your calculations, and don't forget to keep lots of scrap paper on hand.

Make a Commitment

Success does not come without effort. Make the commitment to improve your measurement and conversion skills, and work for understanding. *Why* you do a problem is as important as *how* you do it. If you truly want to be successful, make a commitment to spend the time you need to do a good job. You can do it! When you achieve measurement and conversion success, you have laid the foundation for future challenges and success. So sharpen your pencil and practice!

1

Measurement—Perimeter

This chapter starts with the most basic of the measurements—perimeter. These are problems that involve measuring the distance around a geometric shape. You can think of perimeter as being a boundary for a polygon. Circumference is the name given to the perimeter of a circle. Study the formulas before moving on to working the problems in this section. Remember that perimeter is an addition concept. Read each question carefully and always double check your answer. When you finish, carefully read over the answer explanations.

Formulas

To find the perimeter of any object, add the lengths of all sides of the object.

Square with side length s

$$P = s + s + s + s$$

Rectangle with length l and width w

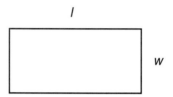

$$P = l + l + w + w$$

Circle with diameter d

Note: The distance around a circle is called the circumference rather than the perimeter.

$$C = \pi d$$

1. Find the perimeter of a rectangle with length 4 meters and width 1 meter.
  a. 5 meters
  b. 6 meters
  c. 8 meters
  d. 10 meters
  e. 15 meters

2. Find the circumference of the circle below.

6 ft

  a. 12π feet
  b. 6π feet
  c. 3π feet
  d. 36π feet
  e. 48π feet

3. Martha's garden is a square. Each side of the garden measures 2 yards. Martha wants to enclose her garden with a fence. How many yards of fence does she need?
  a. 2 yards
  b. 4 yards
  c. 6 yards
  d. 8 yards
  e. 10 yards

4. What is the perimeter of the figure below?

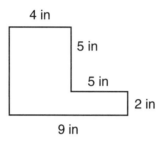

a. 25 inches
b. 32 inches
c. 30 inches
d. 38 inches
e. 40 inches

5. The diameter of a circular swimming pool is 16 feet. What is the circumference of the pool?
a. 50.24 feet
b. 100.48 feet
c. 25.12 feet
d. 200.96 feet
e. 803.84 feet

6. Find the perimeter of the regular hexagon below.

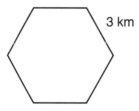

a. 9 km
b. 18 km
c. 24 km
d. 30 km
e. 45 km

7. When measuring the perimeter of a house, what unit would you use to report the results?

    a. centimeters

    b. inches

    c. meters

    d. kilometers

    e. miles

8. Find the perimeter of the figure below.

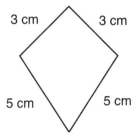

3 cm   3 cm

5 cm   5 cm

    a. 8 cm

    b. 21 cm

    c. 15 cm

    d. 75 cm

    e. 16 cm

9. The perimeter of a square is 44 kilometers. What is the length of one side of the square?

    a. 22 kilometers

    b. 6.6 kilometers

    c. 176 kilometers

    d. 11 kilometers

    e. 88 kilometers

10. Find the circumference of the circle below.

14 yds

   a. 14π yards
   b. 7π yards
   c. 49π yards
   d. 196π yards
   e. 28π yards

11. When measuring the perimeter of a kitchen table, which measurement tool should you use?
   a. scale
   b. tape measure
   c. thermometer
   d. protractor
   e. barometer

12. Find the perimeter of the quadrilateral below.

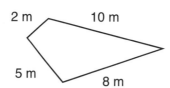

2 m   10 m

5 m   8 m

   a. 20 m
   b. 22 m
   c. 25 m
   d. 27 m
   e. 31 m

13. The perimeter of the figure below is 28 feet. Find the length of the missing side.

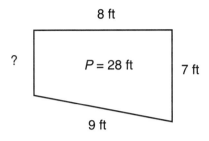

8 ft

? $P = 28$ ft   7 ft

9 ft

a. 4 ft
b. 6 ft
c. 8 ft
d. 10 ft
e. 12 ft

14. What is the perimeter of a baseball diamond if the distance between bases is 90 feet?

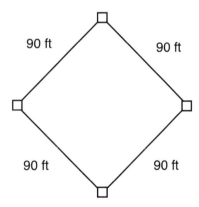

90 ft   90 ft

90 ft   90 ft

a. 22.5 feet
b. 30 feet
c. 90 feet
d. 360 feet
e. 8,100 feet

15. When measuring the perimeter of desk, what unit would you use to report the results?

a. millimeters

b. pounds

c. inches

d. miles

e. kilograms

16. The perimeter of a rectangle is 38 cm. If the length of the rectangle is 12 cm, find the width.

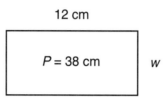

a. 7 cm

b. 14 cm

c. 24 cm

d. 26 cm

e. 18 cm

17. Find the perimeter of the regular pentagon below.

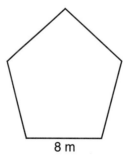

a. 32 m

b. 16 m

c. 20 m

d. 48 m

e. 40 m

18. When measuring the perimeter of a book, which measurement tool should you use?

a. scale

b. protractor

c. ruler

d. thermometer

e. calculator

19. Jason is calculating the amount of fencing he needs to enclose his rectangular garden. The garden measures 8 feet by 10 feet. How much fencing does Jason need to enclose the garden?

a. 18 feet

b. 26 feet

c. 36 feet

d. 64 feet

e. 80 feet

20. Find the circumference of a circle with a radius of 20 miles.

a. 62.8 miles

b. 125.6 miles

c. 1256 miles

d. 1600 miles

e. 31.4 miles

21. Find the perimeter of the shape below.

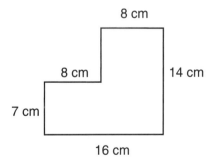

a. 30 cm

b. 29 cm

c. 60 cm

d. 53 cm

e. 44 cm

22. A decorator wants to put lace around a circular pillow. The pillow has a diameter of 14 inches. How much lace does the decorator need?

   a. 21.98 inches
   b. 43.96 inches
   c. 65.94 inches
   d. 153.86 inches
   e. 615.44 inches

23. When measuring the border (perimeter) of the United States, what unit would you use to report the results?

   a. inches
   b. kilograms
   c. meters
   d. tons
   e. miles

24. The perimeter of Kim's rectangular deck is 50 feet. Find the length of the deck if the width is 10 feet.

   a. 15 feet
   b. 40 feet
   c. 60 feet
   d. 10 feet
   e. 20 feet

25. Find the circumference of the circle below.

9 cm

   a. 9π cm
   b. 18π cm
   c. 36π cm
   d. 81π cm
   e. 324π cm

26. Find the distance around the shape below.

12 ft

    a. 30.84 ft
    b. 37.68 ft
    c. 18.84 ft
    d. 113.04 ft
    e. 49.68 ft

27. Find the perimeter of the regular hexagon below with side length 4 kilometers.

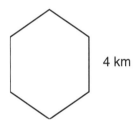

4 km

    a. 8 km
    b. 12 km
    c. 16 km
    d. 20 km
    e. 24 km

28. George walked around the perimeter of Doolittle Park. He kept track of the distance walked on each of the five sides of the park. The first side was 0.5 miles, the second side was 1.0 mile, the third side was 1.5 miles, the fourth side was 0.75 miles, and the fifth side was 1.5 miles. What is the perimeter of the park?
    a. 5.25 miles
    b. 5 miles
    c. 7.25 miles
    d. 4.5 miles
    e. 6.25 miles

29. Find the perimeter of the figure below.

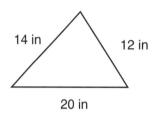

14 in   12 in

20 in

    a. 26 inches
    b. 32 inches
    c. 34 inches
    d. 36 inches
    e. 46 inches

30. Find the circumference of a circle with a radius of 9 inches.
    a. 9π inches
    b. 3π inches
    c. 18π inches
    d. 81π inches
    e. 324π inches

31. A rectangular board has a perimeter of 30 inches. Find the length if the width is 5 inches.
    a. 25 inches
    b. 20 inches
    c. 10 inches
    d. 15 inches
    e. 5 inches

32. When measuring the perimeter of an in-ground pool, what measurement tool should you use?
    a. scale
    b. protractor
    c. ruler
    d. thermometer
    e. tape measure

33. Find the circumference of a circle with a diameter of 22 meters.

    a. 69.08 meters

    b. 138.16 meters

    c. 34.54 meters

    d. 379.94 meters

    e. 245.14 meters

34. Find the perimeter of the equilateral triangle below.

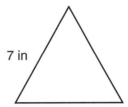

7 in

    a. 14 inches

    b. 49 inches

    c. 28 inches

    d. 21 inches

    e. 53 inches

35. Jenny wants to put a border around her rectangular flower garden. The garden measures 4 feet by 6 feet. Find the perimeter of the garden.

    a. 10 feet

    b. 20 feet

    c. 24 feet

    d. 36 feet

    e. 16 feet

36. Find the perimeter of the figure below.

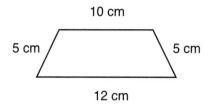

a. 20 cm
b. 22 cm
c. 32 cm
d. 42 cm
e. 50 cm

37. When measuring the perimeter of an office building, what units would you use to report the results?
a. inches
b. gallons
c. tons
d. meters
e. grams

38. Find the perimeter of a rectangle that measures 14 inches by 12 inches.
a. 26 inches
b. 52 inches
c. 144 inches
d. 168 inches
e. 196 inches

39. Find the circumference of a circle with a radius of 1 cm.
a. 3.14 cm
b. 6.28 cm
c. 12.56 cm
d. 9.14 cm
e. 14.34 cm

40. The perimeter of the figure below is 25 inches. Find the length of the missing side.

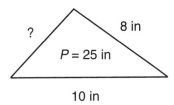

a. 8 inches
b. 7 inches
c. 6 inches
d. 5 inches
e. 4 inches

Answers

1. d. To find the perimeter of a rectangle, find the sum of all four sides; $P = 4 + 4 + 1 + 1$. The perimeter is 10 meters.

2. a. The formula for the circumference of a circle is $C = \pi d$. The radius (half way across the circle) is given. Double the radius to find the diameter. Therefore, the diameter is 12 feet; $C = \pi(12)$ which is equivalent to 12π feet.

3. d. All sides of a square are the same length. Therefore, all four sides of the square are 2 yards. The distance around the square is $2 + 2 + 2 + 2 = 8$ yards.

4. b. Find the sum of all sides of the figure. The left side is not labeled so you must figure out the length of that side before finding the sum. The missing side is the same as the 5 inch side plus the 2 inch side put together (see figure below). Therefore, the length of the side is 7 inches. Next, add all of the sides to find the perimeter; $P = 7 + 4 + 5 + 5 + 2 + 9 = 32$ inches.

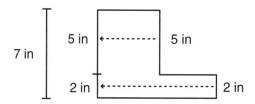

5. a. The formula for circumference is $C = \pi d$. Use 3.14 for π; $C = (3.14)(16) = 50.24$ feet.

6. b. Find the sum of all the sides of the hexagon. The question states that it is a *regular* hexagon which means that all sides are the same length. In other words, each side is 3 km; $P = 3 + 3 + 3 + 3 + 3 + 3 = 18$ km.

7. c. Meters are the appropriate unit for measuring the perimeter of a house. Centimeters and inches are too small and kilometers and miles are too large.

8. e. Find the sum of all the side lengths; $P = 3 + 3 + 5 + 5 = 16$ cm.

9. d. A square has four equal sides. Divide the perimeter by 4 to find the length of a side; $44 \div 4 = 11$ kilometers.

10. a. The formula for circumference is $C = \pi d$. The diameter of the circle is 14 yards; $C = \pi(14)$ which is equivalent to 14π yards.

11. b. A tape measure is used to measure length and is appropriate to use to measure the length of a table. A scale measures weight. A thermometer measures temperature. A protractor measures angles. A barometer measures atmospheric pressure.

12. c. Find the sum of the side lengths; $P = 2 + 10 + 8 + 5 = 25$ m.

13. a. All four sides must add to 28. The three given sides add to 24 feet $(8 + 7 + 9 = 24)$. Therefore, 4 more feet are needed to make 28 feet. The length of the missing side is 4 feet.

14. d. Find the sum of the four sides; $P = 90 + 90 + 90 + 90 = 360$ ft.

15. c. Inches are the appropriate unit to measure the perimeter of a desk. Millimeters are too small. Pounds measure weight. Miles are too large. Kilograms are used to measure weight.

16. a. The formula for the perimeter of a rectangle is $P = l + l + w + w$. Substitute the given information into the equation; $38 = 12 + 12 + w + w$; $38 = 24 + w + w$. In order to have a sum of 38, the widths must add to 14 cm. Since the widths are the same, the width is 7 cm.

17. e. A regular pentagon has five sides that are all the same length. Since all 5 sides have a length of 8 m, the perimeter is $P = 8 + 8 + 8 + 8 + 8 = 40$ m.

18. c. A ruler is the appropriate tool. A scale measures weight. A protractor measures angles. A thermometer measures temperature. A calculator makes mathematical calculations such as addition or multiplication.

19. c. Find the sum of the lengths of all 4 sides of the garden; $P = 10 + 8 + 10 + 8 = 36$ feet.

20. b. The formula for circumference is $C = \pi d$. To find the diameter, double the radius ($20 \times 2 = 40$) Use 3.14 for π; $C = (3.14)(40) = 125.6$ miles.

21. c. Find the sum of all the sides in the figure. One side is missing. To find the missing side, notice that the side length of 7 cm and the missing side add together to make the side length of 14 cm (see figure below). Therefore, the missing side must also be 7 cm; $P = 8 + 7 + 8 + 14 + 16 + 7 = 60$ cm.

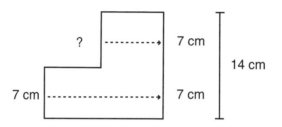

22. b. The formula for circumference is $C = \pi d$. Use 3.14 for π; $C = (3.14)(14) = 43.96$ inches.

23. e. Miles are the appropriate unit. Inches and meters are too small. Kilograms and tons measure weight.

24. a. The formula for the perimeter of a rectangle is $P = l + l + w + w$. The perimeter is 50 feet and the width is 10 feet; $50 = l + l + 10 + 10$. Therefore, the two lengths must add to 30 feet in order to get the perimeter of 50 feet. Since both lengths are the same, each one is 15 feet ($30 \div 2 = 15$).

25. b. The formula for circumference of a circle is $C = \pi d$. The radius is 9 cm. Therefore, the diameter is 18 cm; $C = \pi(18)$ which is equivalent to 18π cm.

26. a. The rounded part of the shape is half of a circle with diameter 12 feet. Find the circumference of a full circle and divide it by two, then add the straight part of the shape which is 12 ft; $C = (3.14)(12) = 37.68$. Half of this circumference is 18.84 ($37.68 \div 2$). Add the straight part to the rounded part; $12 + 18.84 = 30.84$ feet.

27. e. A regular hexagon has six sides of equal length. Therefore, all six sides of the hexagon are 4 km. Find the perimeter by adding all the side lengths together; $P = 4 + 4 + 4 + 4 + 4 + 4 = 24$ km.

28. a. Find the sum of all five sides; $P = 0.5 + 1.0 + 1.5 + 0.75 + 1.5 = 5.25$ miles.

29. e. Find the sum of all three sides; $P = 14 + 12 + 20 = 46$ in.

30. c. The formula for circumference is $C = \pi d$. The radius of the circle is 9 inches. The diameter is double the radius, 18 inches. Therefore, the circumference is 18π inches.

31. c. The formula for the perimeter of a rectangle is $P = l + l + w + w$. Substitute in the given information; $30 = l + l + 5 + 5$. The two lengths must add to 20 to make a perimeter of 30 inches. Since the lengths are equal, each length is 10 inches.

32. e. A tape measure is the appropriate tool. A scale measures weight. A protractor measures angles. A ruler is too small. A thermometer measures temperature.

33. a. The formula for circumference of a circle is $C = \pi d$. Use 3.14 for π; $C = (3.14)(22) = 69.08$ meters.

34. d. All three sides of an equilateral triangle are the same length. Find the perimeter by adding all three side lengths; $P = 7 + 7 + 7 = 21$ inches.

35. b. Find the sum of all four sides of the rectangle; $P = 4 + 4 + 6 + 6 = 20$ feet.

36. c. Find the sum of the four sides of the trapezoid; $P = 5 + 10 + 5 + 12 = 32$ cm.

37. d. Meters are the appropriate unit of measure. Inches are too small. Gallons measure volume. Tons and grams measure weight.

38. b. Find the sum of all four sides of the rectangle; $P = 14 + 14 + 12 + 12 = 52$ inches.

39. b. The formula for circumference is $C = \pi d$. The radius is 1 cm. Double the radius to find the diameter. The diameter is 2 cm. Use 3.14 for π; $C = (3.14)(2) = 6.28$ cm.

40. **b.** The sum of the three sides must be 25 inches. The sum of the two given sides is 18 inches. Therefore, the remaining side must be 7 inches; 7 + 8 + 10 = 25.

2

Measurement— Area

This chapter offers practice in the common measurement of area. Area is the amount of square units that it takes to cover a polygon. Area is a multiplication concept. Generally speaking, area is two measurement quantities that are multiplied together to yield a measurement in square units. Examine the formulas, which includes the area formulas for the basic geometric polygons. You should have these formulas memorized. As always, double-check your answers. The answer explanations that follow the questions will provide one or more correct methods of solution.

Formulas

Area of a triangle

$A = \frac{1}{2}bh$, where b = the base and h = the height

Area of a parallelogram, rectangle, rhombus, or square

$A = bh$, where b = the base and h = the height
OR
$A = lw$, where l = the length and w = the width

Area of a trapezoid

$A = \frac{1}{2}h(b_1 + b_2)$, where h = the height, and b_1 and b_2 = the two parallel bases

Area of a circle

$A = \pi r^2$, where r = the radius

Area of a regular polygon

$A = \frac{1}{2}aP$, where a = length of the apothem (the perpendicular segment drawn from the center of the polygon to one of its sides) and P = the perimeter of the polygon

41. The area of a region is measured in

   a. units.

   b. square units.

   c. cubic units.

   d. quadrants.

   e. none of these

42. When calculating the area of a figure you are finding

   a. the distance around the object.

   b. the length of a side.

   c. the amount of space that the object covers.

   d. the number of sides it has.

   e. the number of degrees in the angles.

43. If the base of a triangle measures 5 m and the height is 4 m, what is the area of the triangle?

   a. 9 m

   b. 10 m

   c. 10 m^2

   d. 20 m^2

   e. 40 m^2

44. What is the area of the figure?

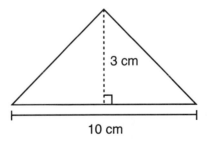

   a. 13 cm^2

   b. 6.5 cm^2

   c. 30 cm^2

   d. 15 cm^2

   e. 60 cm^2

45. What is the area of the figure?

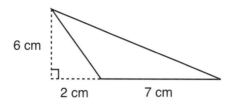

a. 21 cm²
b. 10.5 cm²
c. 27 cm²
d. 42 cm²
e. 54 cm²

46. If the area of the triangle below is 36 in², what is the height?

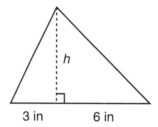

a. 2 in
b. 4 in
c. 6 in
d. 8 in
e. 12 in

47. If the hypotenuse of a right triangle is 13 m and the measure of a leg is 5 m, what is the area of the triangle?
a. 12 m²
b. 30 m²
c. 32.5 m²
d. 60 m²
e. 65 m²

48. If the height of a triangle is 7.2 ft and the base is 5.5 ft, what is the area of the triangle?

   a. 39.6 ft

   b. 19.8 ft²

   c. 79.2 ft²

   d. 12.7 ft²

   e. 39.6 ft²

49. A patio is in the shape of an equilateral triangle. If the length of one side is 4 m, what is the area of the patio to the nearest tenth?

   a. 13.8 m²

   b. 13.9 m²

   c. 6.8 m²

   d. 6.9 m²

   e. none of these

50. What is the area of the figure below?

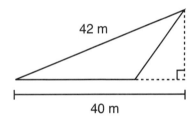

   a. 1,680 m²

   b. 256 m²

   c. 840 m²

   d. 82 m²

   e. cannot be determined

51. Given a triangle with a base of 16 inches and an area of 44 square inches, what is the height?

   a. 2.5 in

   b. 4.5 in

   c. 5.5 in

   d. 9 in

   e. 2.75 in

52. How does the area of a triangle change if the height of the original triangle is doubled?
   a. The area remains the same.
   b. The area is doubled.
   c. The area is tripled.
   d. The area is quadrupled.
   e. The area cannot be determined.

53. If the base of a rectangle is 10 m and the height is 5 m, what is the area of the rectangle?
   a. 15 m^2
   b. 30 m^2
   c. 25 m^2
   d. 50 m^2
   e. 100 m^2

54. The side of a square measures 20 feet. What is the area?
   a. 40 ft
   b. 80 ft^2
   c. 200 ft^2
   d. 2,000 ft^2
   e. 400 ft^2

55. The base of a parallelogram is 25 m and the height is 15 m. What is the area?
   a. 40 m^2
   b. 80 m^2
   c. 375 m^2
   d. 187.5 m^2
   e. 750 m^2

56. What is the area of the figure below?

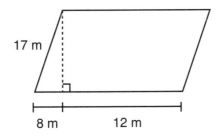

a. 300 m²

b. 340 m²

c. 136 m²

d. 204 m²

e. 180 m²

57. How does the area of a rectangle change if both the base and the height of the original rectangle are tripled?

a. The area is tripled.

b. The area is six times larger.

c. The area is nine times larger.

d. The area remains the same.

e. The area cannot be determined.

58. What is the area of the figure below?

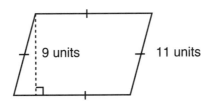

a. 20 square units

b. 99 square units

c. 22 square units

d. 77 square units

e. 121 square units

59. A rhombus has a side length of 30 cm. What is the area of the rhombus?
   a. 30 cm^2
   b. 60 cm^2
   c. 120 cm^2
   d. 900 cm^2
   e. cannot be determined

60. The area of a rectangular yard is 504 feet^2. If the length of the yard is 21 feet, what is the width?
   a. 22 ft
   b. 24 ft
   c. 525 ft
   d. 483 ft
   e. 10,584 ft

61. What is the area of the parallelogram below?

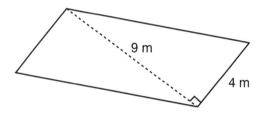

   a. 13 m^2
   b. 18 m^2
   c. 27 m^2
   d. 36 m^2
   e. cannot be determined

62. Brooke wants to carpet her rectangular bedroom. If the length of the room is 10 feet and the width is 12 feet, how much will the cost be if the carpet costs $2.94 per square foot?
   a. $29.40
   b. $35.28
   c. $352.80
   d. $64.68
   e. $423.36

63. What is the area of the trapezoid below?

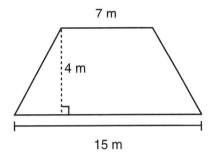

a. 22 m²
b. 28 m²
c. 44 m²
d. 60 m²
e. 210 m²

64. Cathy has a countertop in the shape of a right trapezoid that she would like to paint. If the dimensions of the countertop are as shown in the figure below, what is the total area she will paint?

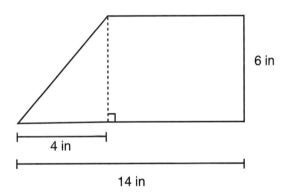

a. 420 in²
b. 42 in²
c. 56 in²
d. 84 in²
e. 72 in²

65. If the area of a trapezoid is 112.5 cm² and the two parallel bases measure 23 cm and 27 cm, what is the height of the trapezoid?

   a. 62.5 cm

   b. 25 cm

   c. 2.5 cm

   d. 5 cm

   e. 4.5 cm

66. Given the following figure, what is the area of the trapezoid?

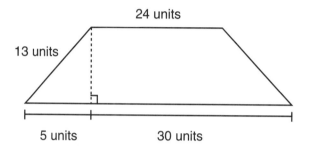

   a. 59 square units

   b. 12 square units

   c. 383.5 square units

   d. 354 square units

   e. 455 square units

67. What is the length of a base of a trapezoid if the other parallel base is 10 m, the height is 3.4 m, and the area is 28.9 m²?

   a. 7 m

   b. 8.5 m

   c. 2.8 m

   d. 15.5 m

   e. 2.9 m

68. When measuring the area of a football field, you would most likely use

   a. square inches.

   b. square millimeters.

   c. square miles.

   d. square yards.

   e. square liters.

69. What is the area of a regular pentagon with a side length of 6 cm and an apothem length of 5 cm?

   a. 15 cm²

   b. 30 cm²

   c. 60 cm²

   d. 75 cm²

   e. 150 cm²

70. A tile in the shape of a regular hexagon has a side length of 3 in and an apothem length of 4 in. What is the area of the tile?

   a. 12 in²

   b. 18 in²

   c. 36 in²

   d. 72 in²

   e. 144 in²

71. A stop sign in the shape of a regular octagon has an area of 160 cm². If the length of the apothem is 10 cm, how long is one side of the sign?

   a. 2 cm

   b. 4 cm

   c. 8 cm

   d. 16 cm

   e. 32 cm

72. If the circle below has a radius of 10 m, what is the area of the circle?

   a. 10 m²

   b. 10π m²

   c. 20π m²

   d. 100 m²

   e. 100π m²

73. If the diameter of a circle is 11 inches, what is the area of the
circle, to the nearest tenth? Use $\pi = 3.14$.
 a. 379.9 in²
 b. 95.0 in²
 c. 30.3 in²
 d. 72 in²
 e. 94.9 in²

74. If the area of a circle is 42.25π m², what is the length of the radius
of the circle?
 a. 10.5 m
 b. 6.5 m
 c. 7.5 m
 d. 21.125 m
 e. 2.54 m

75. If the area of a circle is 254.34 in², what is the length of the
diameter of the circle? Use $\pi = 3.14$.
 a. 3 in
 b. 9 in
 c. 18 in
 d. 27 in
 e. none of these

76. What is the area of the figure below if the diameter of the circle is
16 centimeters?

 a. 64 cm²
 b. 256π cm²
 c. 16π cm²
 d. 64π cm²
 e. 256 cm²

77. At the Pizza Place, the diameter of a small round pizza is 9 inches and the diameter of a large round pizza is 15 inches. Approximately how much more pizza do you get in a large than a small pizza? Use $\pi = 3.14$.

    a. 113 in²

    b. 144 in²

    c. 6 in²

    d. 19 in²

    e. 452 in²

78. While waiting to go for a walk, a dog on a leash is tied to a stake in the ground. If the leash is 7 feet long, what is the approximate area that the dog can roam while tied to the stake? Use $\pi = 3.14$.

    a. 44 ft²

    b. 615 ft²

    c. 36 ft²

    d. 49 ft²

    e. 154 ft²

79. What is the area of the irregular figure below? Use $\pi = 3.14$.

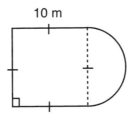

10 m

    a. 139.25 m²

    b. 178.5 m²

    c. 78.5 m²

    d. 414 m²

    e. 109.5 m²

80. What is the area of the irregular figure below?

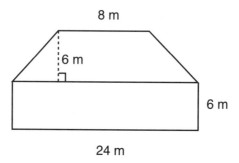

8 m

6 m

6 m

24 m

a. 49 m²
b. 144 m²
c. 240 m²
d. 192 m²
e. 336 m²

81. What is the area of the irregular figure below? Use π = 3.14.

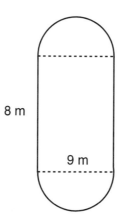

8 m

9 m

a. 135.585 m²
b. 326.34 m²
c. 103.79 m²
d. 127.585 m²
e. 144.585 m²

82. What is the area of the shaded region?

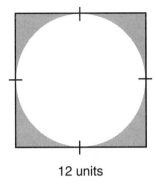

12 units

    a. $(36\pi - 144)$ square units

    b. $(36\pi + 144)$ square units

    c. 144 square units

    d. $(144 - 144\pi)$ square units

    e. $(144 - 36\pi)$ square units

83. What is the area of the shaded region?

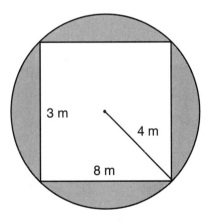

3 m

4 m

8 m

    a. $(64\pi - 12)$ m^2

    b. $(16\pi - 24)$ m^2

    c. $(16\pi - 12)$ m^2

    d. $(24 - 16\pi)$ m^2

    e. $(24 - 64\pi)$ m^2

84. A diagram of a washer is drawn in the figure below. If the shaded region represents the washer, what is its area? Use π = 3.14.

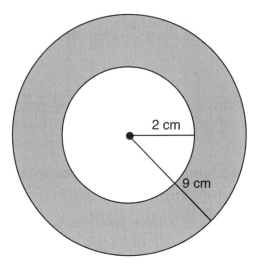

    a. 21.98 cm²

    b. 254.34 cm²

    c. 241.78 cm²

    d. 248.06 cm²

    e. none of these

85. What is the area of the shaded region?

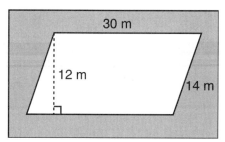

    a. 444 m²

    b. 696 m²

    c. 78 m²

    d. 4,176 m²

    e. 504 m²

Answers

41. b. When calculating area you are finding the number of square units that cover the region.

42. c. The area of a figure is the amount of space the object covers, in square units. The distance around a figure is called the *perimeter* of the figure.

43. c. The area of a triangle can be found by using the formula $A = \frac{1}{2}bh$, where b is the base of the triangle and h is the height. Substitute into the formula to get $A = \frac{1}{2}(5)(4) = \frac{1}{2}(20) = 10$ m². Be careful to select the answer choice with the correct units, which should be square meters (m²).

44. d. The area of a triangle can be found by using the formula $A = \frac{1}{2}bh$, where b is the base of the triangle and h is the height. Substitute into the formula to get $A = \frac{1}{2}(10)(3) = \frac{1}{2}(30) = 15$ cm².

45. a. The area of a triangle can be found by using the formula $A = \frac{1}{2}bh$, where b is the base of the triangle and h is the height. Substitute into the formula to get $A = \frac{1}{2}(6)(7) = \frac{1}{2}(42) = 21$ cm². Be aware that the height of the triangle may be located outside of the figure, as it was in this problem. The length of the base of this triangle is only 7 cm, not 9 cm.

46. d. Use the formula $A = \frac{1}{2}bh$, where b is the base of the triangle and h is the height. Since the area is given and the base is 9 in, substitute into the formula to get $36 = \frac{1}{2}(9)(h)$. Multiply each side of the equation by 2 to eliminate the fraction; $72 = 9h$. Divide both sides of the equation by 9 to get $h = 8$ in.

47. b. First, use the Pythagorean theorem ($a^2 + b^2 = c^2$) to find the missing side of the right triangle, which is the base side; $5^2 + b^2 = 13^2$; $25 + b^2 = 169$. Subtract 25 from both sides of the equation to get $b^2 = 144$. So, $b = 12$. The area of a triangle can be found by using the formula $A = \frac{1}{2}bh$, where b is the base of the triangle and h is the height. Note that the height is also a side because it is a right triangle. Substitute into the formula to get $A = \frac{1}{2}(12)(5) = \frac{1}{2}(60) = 30$ m².

48. **b.** The area of a triangle can be found by using the formula $A = \frac{1}{2}bh$, where b is the base of the triangle and h is the height. Substitute into the formula to get $A = \frac{1}{2}(5.5)(7.2) = \frac{1}{2}(39.6) = 19.8$ ft^2.

49. **d.** First, use the Pythagorean theorem ($a^2 + b^2 = c^2$) to find the height of the triangle, which divides the triangle into two congruent right triangles. Use b as the height, $a = 2$ (which is half of the base side and a leg of a right triangle), and $c = 4$ (which is the hypotenuse of a right triangle); $2^2 + b^2 = 4^2$; $4 + b^2 = 16$. Subtract 4 from both sides of the equation to get $b^2 = 12$. So, $b \approx 3.464$. Then, the area of the triangle can be found by using the formula $A = \frac{1}{2}bh$, where b is the base of the equilateral triangle and h is the height. Substitute into the formula to get $A = \frac{1}{2}(4)(3.464) = \frac{1}{2}(13.856) = 6.928 \approx 6.9$ m^2.

50. **e.** The height of the triangle needs to be perpendicular to the base. Since the height is not known, the area of this triangle cannot be determined.

51. **c.** Use the formula $A = \frac{1}{2}bh$, where b is the base of the triangle and h is the height. Since the area is given and the base is 16 in, substitute into the formula to get $44 = \frac{1}{2}(16)(h)$. Multiply each side of the equation by 2 to eliminate the fraction; $88 = 16h$. Divide both sides of the equation by 16 to get $h = 5.5$ inches.

52. **b.** Since one dimension is doubled, there is an additional factor of 2. Therefore the new area is twice as large as the original, or doubled. For example, use a triangle with a base of 10 and height of 4. The area is $\frac{1}{2}$ of 40, which is 20 square units. If you multiply the side length 10 by 2, the new dimensions are 20 and 4. The new area is $\frac{1}{2}$ of 80, which is 40 square units. By comparing the new area with the original area, 40 square units is double the size of 20 square units.

53. **d.** The area of a rectangle can be found by the formula $A = \frac{1}{2}bh$, where b is the base of the rectangle and h is the height. Substitute into the formula to get $A = (10)(5) = 50$ m^2.

54. e. The area of a square can be found by the formula $A = bh$, where b is the base of the square and h is the height. Substitute into the formula to get $A = (20)(20) = 400$ ft^2. Note that the base and height are always the same measure in a square.

55. c. The area of a parallelogram can be found by the formula $A = bh$, where b is the base of the parallelogram and h is the height. Substitute into the formula to get $A = (25)(15) = 375$ m^2.

56. a. The area of a parallelogram can be found by the formula $A = bh$, where b is the base of the parallelogram and h is the height. First, use the right triangle on the left of the figure and the Pythagorean theorem ($a^2 + b^2 = c^2$) to find the height. Use b as the height, $a = 8$, and $c = 17$; $8^2 + b^2 = 17^2$; $64 + b^2 = 289$. Subtract 64 from both sides of the equation to get $b^2 = 225$. So, $b = \sqrt{225} = 15$. Also note that the base is 8 m + 12 m = 20 m. Substitute into the formula to get $A = (20)(15) = 300$ m^2.

57. c. Since both dimensions are tripled, there are two additional factors of 3. Therefore the new area is $3 \times 3 = 9$ times as large as the original. For example, use a rectangle with a base of 5 and height of 6. The area is $5 \times 6 = 30$ square units. If you multiply the each side length by 3, the new dimensions are 15 and 18. The new area is 15×18, which is 270 square units. By comparing the new area with the original area, 270 square units is nine times larger than 30 square units; $30 \times 9 = 270$.

58. b. The area of a rhombus can be found by the formula $A = bh$, where b is the base of the rhombus and h is the height. Note that a rhombus has four equal sides, so the base is 11 units and the height as shown is 9 units. Substitute into the formula to get $A = (11)(9) = 99$ square units.

59. e. Although each side of the rhombus is 30 cm, the height is not known. Therefore, the area cannot be determined. It cannot be assumed that the rhombus is a square and that the consecutive sides are perpendicular.

60. b. Use the formula $A = lw$, where l is the length of the rectangular yard and w is the width. Since the area and length are given, substitute into the formula to get $504 = 21w$. Divide each side of the equation by 21 to get $w = 24$ ft.

61. **d.** The area of a parallelogram can be found by the formula $A = bh$, where b is the base of the parallelogram and h is the height. Note that the since the two known dimensions are perpendicular, 4 m can be used as the base and 9 m as the height. Substitute into the formula to get $A = (4)(9) = 36$ m^2.

62. **c.** First, find the area of the room using the formula $A = lw$, where l is the length of the room and w is the width. Substitute into the formula to get $A = (10)(12) = 120$ ft^2. Multiply the total area by \$2.94 to get the cost of the carpet; $120 \times 2.94 = \$352.80$.

63. **c.** The area of a trapezoid can be found by using the formula $A = \frac{1}{2}h(b_1 + b_2)$, where h is the height of the trapezoid and b_1 and b_2 are the two parallel bases. Substitute into the formula to get $A = \frac{1}{2}(4)(7 + 15) = \frac{1}{2}(4)(22) = \frac{1}{2}(88) = 44$ m^2.

64. **e.** The area of a trapezoid can be found by using the formula $A = \frac{1}{2}h(b_1 + b_2)$, where h is the height of the trapezoid and b_1 and b_2 are the two parallel bases. In this case, the height is 6 inches and the parallel bases are 14 inches and 10 inches ($14 - 4 = 10$). Substitute into the formula to get $A = \frac{1}{2}(6)(14 + 10) = \frac{1}{2}(6)(24) = \frac{1}{2}(144) = 72$ in^2.

65. **e.** Use the formula $A = \frac{1}{2}h(b_1 + b_2)$, where h is the height of the trapezoid and b_1 and b_2 are the two parallel bases. Since the area and the two bases are given, substitute into the formula to get $112.5 = \frac{1}{2}(h)(23 + 27)$. Simplify; $112.5 = \frac{1}{2}(h)(50)$; $112.5 = 25h$. Divide each side of the equation by 25 to get $h = 4.5$ cm.

66. **d.** The area of a trapezoid can be found by using the formula $A = \frac{1}{2}h(b_1 + b_2)$, where h is the height of the trapezoid and b_1 and b_2 are the two parallel bases. First, use the right triangle on the left of the figure and the Pythagorean theorem ($a^2 + b^2 = c^2$) to find the height. Use b as the height, $a = 5$, and $c = 13$; $5^2 + b^2 = 13^2$; $25 + b^2 = 169$. Subtract 25 from both sides of the equation to get $b^2 = 144$. So, $b = \sqrt{144} = 12$. Note that the lengths of the parallel bases are 24 units and 35 units. Substitute into the formula to get $A = \frac{1}{2}(12)(24 + 35) = \frac{1}{2}(12)(59) = \frac{1}{2}(708) = 354$ square units.

67. **a.** Use the formula $A = \frac{1}{2}h(b_1 + b_2)$, where h is the height of the trapezoid and b_1 and b_2 are the two parallel bases. Since the area, the height and one of the bases are given, substitute into the formula to get $28.9 = \frac{1}{2}(3.4)(10 + b_2)$. Simplify; $28.9 = 1.7(10 + b_2)$. Divide each side of the equation by 1.7; $17 = 10 + b_2$. Subtract 10 from each side of the equal sign to get $b_2 = 7$.

68. **d.** A football field would most likely be measured in square yards. Square inches and square millimeters are too small, and square miles are too large. Square liters do not make sense since liters are a measure of capacity, not length or area.

69. **d.** Use the formula $A = \frac{1}{2}aP$, where a = length of the apothem and P = the perimeter of the polygon. Since the figure is a pentagon and has five sides, the perimeter is $6 \times 5 = 30$ cm. Substitute into the formula; $A = \frac{1}{2}(5)(30) = \frac{1}{2}(150) = 75$ cm^2.

70. **c.** Use the formula $A = \frac{1}{2}aP$, where a = length of the apothem and P = the perimeter of the polygon. Since the figure is a hexagon and has six sides, the perimeter is $3 \times 6 = 18$ in. Substitute into the formula; $A = \frac{1}{2}(4)(18) = \frac{1}{2}(72) = 36$ in^2.

71. **b.** Use the formula $A = \frac{1}{2}aP$, where a = length of the apothem and P = the perimeter of the polygon. Since the area and the length of the apothem are given, substitute into the formula to find the perimeter of the octagon; $160 = \frac{1}{2}(10)P$. Simplify; $160 = 5P$. Divide each side of the equal sign by 5 to get $P = 32$. Since an octagon has eight sides, $32 \div 8 = 4$ cm for each side of the sign.

72. **e.** The area of a circle can be found by using the formula $A = \pi r^2$ where r is the radius of the circle. Substitute into the formula; $A = \pi(10)^2 = 100\pi$ m^2.

73. **b.** The area of a circle can be found by using the formula $A = \pi r^2$ where r is the radius of the circle. If the diameter is 11 inches, then the radius is half that, or 5.5 inches. Substitute into the formula; $A = (3.14)(5.5)^2 = (3.14)(30.25) = 94.985 \approx 95.0$ in^2.

74. **b.** Use the formula $A = \pi r^2$ where r is the radius of the circle. Since the area is given, substitute into the formula to find the radius; $42.25\pi = \pi r^2$. Divide each side of the equation by π to get $42.25 = r^2$, so $r = \sqrt{42.25} = 6.5$ m.

75. **c.** Use the formula $A = \pi r^2$ where r is the radius of the circle. Since the area is given, first substitute into the formula to find the radius. Then double the length of the radius to find the diameter; $254.34 = (3.14)r^2$. Divide each side of the equation by 3.14 to get $81 = r^2$, so $r = \sqrt{81} = 9$ in. Therefore, the diameter is 18 inches.

76. **d.** The area of a circle can be found by using the formula $A = \pi r^2$ where r is the radius of the circle. If the diameter is 16 cm, then the radius is half that, or 8 cm. Substitute into the formula; $A = \pi(8)^2 = 64\pi$ cm^2.

77. **a.** To find the difference, find the area of the small pizza and subtract it from the area of the large pizza. Be sure to use the radius of each pizza in the formula, which is half of the diameter; $A_{\text{large}} - A_{\text{small}} = \pi r^2 - \pi r^2 = (3.14)(7.5)^2 - (3.14)(4.5)^2$. Evaluate exponents; $(3.14)(56.25) - (3.14)(20.25)$. Multiply within each term; $176.625 - 63.585$. Subtract to get 113.04 in^2. There are approximately 113 more square inches in the large pizza.

78. **e.** The area in which the dog can roam is a circle with the center at the stake and a radius that is the length of the leash. Use the formula $A = \pi r^2$ where r is the radius of the circle; $A = (3.14)(7)^2 = (3.14)(49) = 153.86$. The dog has about 154 square feet to roam in.

79. **a.** The figure is made up of a square and a half-circle. The square has sides that are 10 m and the diameter of the semicircle is a side of the square. Therefore, the radius of the circle is 5 m. To find the area of the figure, find the area of each part and add the areas together; $A_{\text{square}} + A_{\frac{1}{2}\text{ circle}} = bh + \frac{1}{2}\pi r^2 = (10)(10) + \frac{1}{2}(3.14)(5)^2$. Simplify the expression; $100 + \frac{1}{2}(3.14)(25) = 100 + \frac{1}{2}(78.5) = 100 + 39.25$. Therefore the total area is 139.25 m^2.

80. **c.** The figure is made up of a rectangle and trapezoid. Find the area of each section and then add the areas together to find the total area of the figure; $A_{\text{rectangle}} + A_{\text{trapezoid}} = bh + \frac{1}{2}h(b_1 + b_2) = (24)(6) + \frac{1}{2}(6)(24 + 8)$. Simplify the expression; $144 + \frac{1}{2}(6)(32) = 144 + \frac{1}{2}(192) = 144 + 96$. Therefore, the total area is 240 m^2.

81. **a.** The figure is made up of a rectangle and two half-circles of the same radius. Find the area of the rectangle and treat the two half-circles as a whole circle to find the total area of the figure. Note that the radius of the circles is one-half of the width of the rectangle, or 4.5 m; $A_{rectangle} + A_{circle} = bh + \pi r^2 = (8)(9) + (3.14)(4.5)^2$. Simplify by evaluating exponents first and then multiplying within each term; $(8)(9) + (3.14)(20.25) = 72 + 63.585$. This combines for a total area of 135.585 m^2.

82. **e.** The area of the shaded region is the area of the inner figure subtracted from the area of the outer figure; $A_{shaded} = A_{outer} - A_{inner}$. In this figure, the outer region is a square and the inner region is a circle; $A_{shaded} = A_{square} - A_{circle} = bh - \pi r^2$. Substitute into the formula. Since the circle is inscribed in the square, the radius of the circle is one-half the length of a side of the square; $(12)(12) - \pi(6)^2 = (144 - 36\pi)$ sq. units.

83. **b.** The area of the shaded region is the area of the inner figure subtracted from the area of the outer figure; $A_{shaded} = A_{outer} - A_{inner}$. In this figure, the outer region is a circle and the inner region is a parallelogram; $A_{shaded} = A_{circle} - A_{parallelogram} = \pi r^2 - bh$. Substitute into the formula; $\pi(4)^2 - (3)(8) = (16\pi - 24)$m^2.

84. **c.** The area of the shaded region is the area of the inner figure subtracted from the area of the outer figure; $A_{shaded} = A_{outer} - A_{inner}$. In this figure, the outer region is the circle with radius 9 cm and the inner region is the circle with radius 2 cm; $A_{shaded} = A_{outercircle} - A_{innercircle} = \pi r^2 - \pi r^2$. Substitute into the formula; $\pi(9)^2 - \pi(2)^2 = 81\pi - 4\pi = 77\pi$; $77(3.14) = 241.78$ cm^2.

85. **e.** The area of the shaded region is the area of the inner figure subtracted from the area of the outer figure; $A_{shaded} = A_{outer} - A_{inner}$. In this figure, the outer region is a rectangle and the inner region is a parallelogram; $A_{shaded} = A_{rectangle} - A_{parallelogram} = bh - bh$. Substitute into the formula and multiply before subtracting; $(36)(24) - (30)(12) = 864 - 360 = 504$ m^2.

3

Measurement— Volume

This chapter contains a series of 45 questions that review another common measurement—volume. These are problems that involve measuring the amount of cubic units that it takes to *fill* a three-dimensional solid. Like area, volume is also a multiplication concept. Volume formulas are based, generally speaking, on *three* measurement quantities that are multiplied together to yield a measurement in *cubic* units. Study the formulas, which include the volume formulas for the basic geometric solids. For most questions, you will be expected to have these formulas memorized. Always double-check your solution. Often on multiple-choice tests, common errors that are made by a majority of people will be offered as choices. The answer explanations provide a correct method of solution. Read over the explanations for all problems, regardless of whether you answered the question correctly or not. A possible alternate method of solution may be presented that will further your knowledge and skill at solving measurement problems.

Formulas

Volume of a rectangular prism

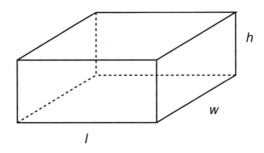

$$V = l \times w \times h$$

Volume of a cone

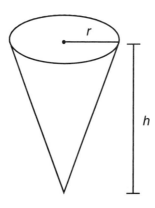

$$V = \tfrac{1}{3}\pi r^2 h$$

Volume of a cylinder

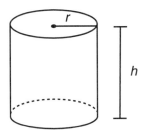

$$V = \pi r^2 h$$

Volume of a sphere

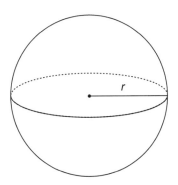

$$V = \frac{4}{3}\pi r^3$$

Volume of a pyramid

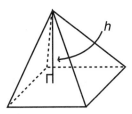

$$V = \frac{1}{3} \times area\ of\ base \times height$$

86. A can containing orange juice concentrate has a diameter of 6 cm and a height of 12 cm. Find the volume of the can.
  a. 36π cm cubed
  b. 72π cm cubed
  c. 432π cm cubed
  d. 108π cm cubed
  e. 256π cm cubed

87. A cardboard box measures $2\frac{1}{2}$ ft by 3 ft by 3 ft. Find the volume of the box.
  a. $8\frac{1}{2}$ ft cubed
  b. 27 ft cubed
  c. $7\frac{1}{2}$ ft cubed
  d. $22\frac{1}{2}$ ft cubed
  e. 25 ft cubed

88. What is the volume of the figure below?

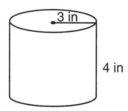

  a. 36π inches cubed
  b. 12π inches cubed
  c. 48π inches cubed
  d. 96π inches cubed
  e. 108π inches cubed

89. When measuring the volume of a box used to package a refrigerator, what unit would you use to report the results?
  a. cubic cm
  b. cubic inches
  c. cubic feet
  d. cubic miles
  e. cubic mm

90. The trailer of a truck is a rectangular prism that measures 20 feet by 8 feet by 10 feet. Find the volume of the trailer.
   a. 1,600 ft cubed
   b. 160 ft cubed
   c. 80 ft cubed
   d. 38 ft cubed
   e. 200 ft cubed

91. What is the volume of a sphere with radius 6 inches? Use 3.14 for π.
   a. 1,296 inches cubed
   b. 904.32 inches cubed
   c. 678.24 inches cubed
   d. 216 inches cubed
   e. 150.72 inches cubed

92. An ice cream cone has a diameter of 4 cm and a height of 10 cm. What is the volume of the cone, rounded to the nearest hundredth? Use 3.14 for π.
   a. 167.47 cm cubed
   b. 82.46 cm cubed
   c. 376.83 cm cubed
   d. 14.27 cm cubed
   e. 41.87 cm cubed

93. What is the volume of the figure below?

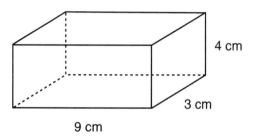

a. 12 cm cubed
b. 27 cm cubed
c. 108 cm cubed
d. 324 cm cubed
e. 801 cm cubed

94. A round swimming pool has a diameter of 12 feet and a height of 4 feet in all locations. Find the volume of the pool.
a. 24π ft cubed
b. 48π ft cubed
c. 98π ft cubed
d. 144π ft cubed
e. 96π ft cubed

95. What is the volume of the figure below, rounded to the nearest hundredth? Use 3.14 for π.

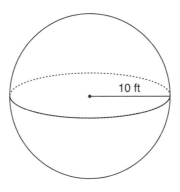

10 ft

a. 4,186.67 ft cubed
b. 133.33 ft cubed
c. 1,256 ft cubed
d. 750 ft cubed
e. 6,325.67 ft cubed

96. What is the volume of the figure below, rounded to the nearest hundredth?

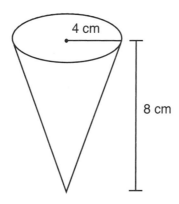

4 cm

8 cm

a. 100.48 cm cubed
b. 133.97 cm cubed
c. 298.76 cm cubed
d. 401.92 cm cubed
e. 98.46 cm cubed

97. A cylindrical fuel tank measures 6 m high and has a diameter of
3 m. Find the volume of the fuel tank.

    a. 9π m cubed

    b. 18π m cubed

    c. 54π m cubed

    d. 13.5π m cubed

    e. 41.5π m cubed

98. What is the volume of the figure below?

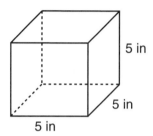

    a. 25 in cubed

    b. 15 in cubed

    c. 125 in cubed

    d. 625 in cubed

    e. 75 in cubed

99. The length of a cereal box is 10 inches, the width is 3 inches, and
the height is 14 inches. What is the volume of the cereal box?

    a. 420 in cubed

    b. 27 in cubed

    c. 1,260 in cubed

    d. 46 in cubed

    e. 340 in cubed

100. A pyramid in Egypt has a square base that measures 45 feet on
each side and a height of 80 feet. Find the volume of the pyramid.

    a. 125 ft cubed

    b. 54,000 ft cubed

    c. 1,200 ft cubed

    d. 96,000 ft cubed

    e. 162,000 ft cubed

101. A funnel is shaped like a cone. The diameter of the funnel's base is 6 inches. The height of the funnel is 6 inches. Find the volume of the funnel.

a. 6π in cubed

b. 72π in cubed

c. 10π in cubed

d. 16π in cubed

e. 18π in cubed

102. What is the volume of the sphere below?

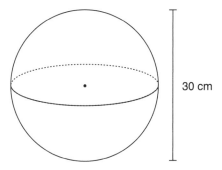

30 cm

a. $3,375\pi$ cm cubed

b. $36,000\pi$ cm cubed

c. $4,500\pi$ cm cubed

d. 333.3π cm cubed

e. 239π cm cubed

103. What is the volume of the cone below, rounded to the nearest hundredth?

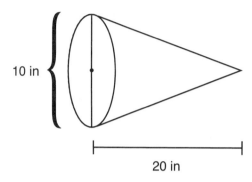

10 in

20 in

    a. 25π in cubed
    b. 76.67π in cubed
    c. 96π in cubed
    d. 166.67π in cubed
    e. 256π in cubed

104. The radius of a ball is 7 inches. Find the volume of the ball, rounded to the units (ones) place.
    a. 457π in cubed
    b. 49π in cubed
    c. 546π in cubed
    d. 786π in cubed
    e. 343π in cubed

105. What is the volume of the cylinder below?

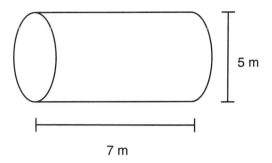

5 m

7 m

   a. 76.75π m cubed

   b. 43.75π m cubed

   c. 98.5π m cubed

   d. 23.5π m cubed

   e. 145.75π m cubed

106. The radius of a basketball is 12 cm. Find the volume of the ball.

   a. 1,728π cm cubed

   b. 2,304π cm cubed

   c. 1,896π cm cubed

   d. 2,104π cm cubed

   e. 340π cm cubed

107. What is the volume of the rectangular prism below?

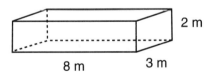

2 m

8 m   3 m

   a. 16 m cubed

   b. 64 m cubed

   c. 32 m cubed

   d. 128 m cubed

   e. 8 m cubed

108. A silo is a cylinder shape. The height of the silo is 35 feet and the diameter of the silo is 20 feet. Find the volume of the silo.
   a. 14,000π ft cubed
   b. 700π ft cubed
   c. 100π ft cubed
   d. 350π ft cubed
   e. 3,500π ft cubed

109. Find the volume of the pyramid below.

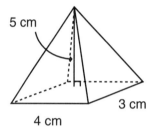

   a. 20 cm cubed
   b. 60 cm cubed
   c. 15 cm cubed
   d. 12 cm cubed
   e. 180 cm cubed

110. Find the volume of the hemisphere below.

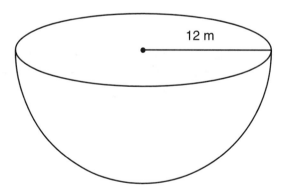

   a. 2,304π m cubed
   b. 3,456π m cubed
   c. 1,152π m cubed
   d. 9,600π m cubed
   e. 1,200π m cubed

111. Find the volume of the shaded region.

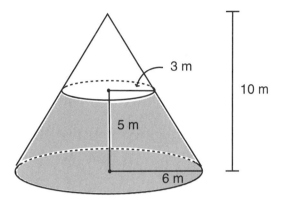

    a. 120π m cubed

    b. 105π m cubed

    c. 15π m cubed

    d. 25π m cubed

    e. 108 m cubed

112. A packing box measures 6 inches high, 10 inches wide, and 5 inches long. Find the volume of the box.

    a. 30 in cubed

    b. 50 in cubed

    c. 60 in cubed

    d. 300 in cubed

    e. 600 in cubed

113. A piece of wood is a perfect cube. Each side of the cube measures 8 cm. Find the volume of the cube.

    a. 64 cm cubed

    b. 81 cm cubed

    c. 729 cm cubed

    d. 24 cm cubed

    e. 512 cm cubed

114. When measuring the volume of a car, what unit would you use to report the results?

    a. miles cubed

    b. mm cubed

    c. feet cubed

    d. km cubed

    e. cm cubed

115. Find the volume of the sphere below.

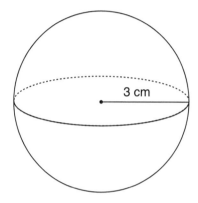

3 cm

    a. 36π cm cubed

    b. 9π cm cubed

    c. 27π cm cubed

    d. 108π cm cubed

    e. 208π cm cubed

116. Find the volume of the cylinder below.

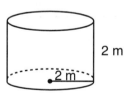

2 m

2 m

    a. 4π m cubed

    b. 6π m cubed

    c. 12π m cubed

    d. 8π m cubed

    e. 16π m cubed

117. A can of soup measures 4 inches high and has a diameter of 2
inches. Find the volume of the can.

    a. 16π in cubed

    b. 4π in cubed

    c. 8π in cubed

    d. 32π in cubed

    e. 10π in cubed

118. Find the volume of the pyramid below.

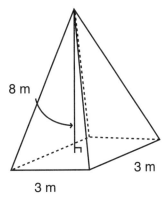

8 m

3 m

3 m

    a. 72 m cubed

    b. 36 m cubed

    c. 24 m cubed

    d. 84 m cubed

    e. 96 m cubed

119. The globe below has a diameter of 14 inches. Find the volume of the globe. Use 3.14 for π.

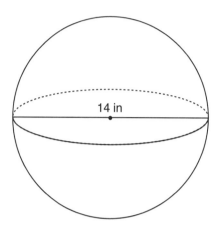

14 in

    a. 65.33 in cubed
    b. 84 in cubed
    c. 1,654.27 in cubed
    d. 9,672.17 in cubed
    e. 1,436.03 in cubed

120. Find the volume of the cone below.

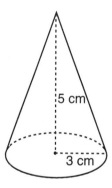

5 cm

3 cm

    a. 15π cm cubed
    b. 45π cm cubed
    c. 81π cm cubed
    d. 12π cm cubed
    e. 21π cm cubed

121. Find the volume of the figure below. Use 3.14 for π. Round your answer to the nearest hundredth.

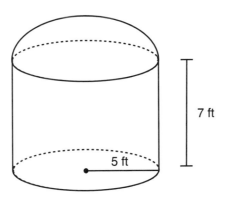

a. 549.5 ft cubed

b. 811.17 ft cubed

c. 261.67 ft cubed

d. 891.27 ft cubed

e. 781.67 ft cubed

122. Find the volume of the figure below. Use 3.14 for π. Round your answer to the nearest hundredth.

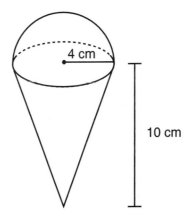

a. 167.47 cm cubed

b. 301.44 cm cubed

c. 187.94 cm cubed

d. 133.97 cm cubed

e. 245.78 cm cubed

123. A cylindrical cement column is 10 feet tall with a radius of 2 feet. Find the volume of the cement column.

    a. 40π ft cubed

    b. 16π ft cubed

    c. 100π ft cubed

    d. 20π ft cubed

    e. 80π ft cubed

124. Find the volume of the figure below.

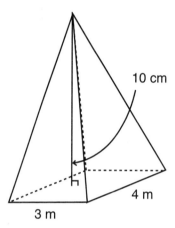

    a. 12 m cubed

    b. 36 m cubed

    c. 39 m cubed

    d. 40 m cubed

    e. 80 m cubed

125. Find the volume of the figure below.

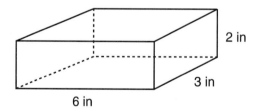

    a. 18 in cubed

    b. 36 in cubed

    c. 72 in cubed

    d. 42 in cubed

    e. 12 in cubed

126. A cabinet is a rectangular prism. Find the volume of the cabinet if it measures 18 inches by 24 inches by 15 inches.
   a. 57 in cubed
   b. 96 in cubed
   c. 6,480 in cubed
   d. 10,246 in cubed
   e. 8,246 in cubed

127. Find the volume of the figure below.

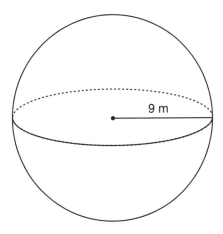

   a. 972π m cubed
   b. 108π m cubed
   c. 324π m cubed
   d. 627π m cubed
   e. 936π m cubed

128. Find the volume of the figure below.

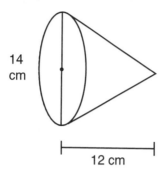

14
cm

12 cm

   a. 49π cm cubed

   b. 144π cm cubed

   c. 184π cm cubed

   d. 196π cm cubed

   e. 218π cm cubed

129. A camera came packed in a 5-inch by 4-inch by 4-inch box. Find the volume of the box.

   a. 80 in cubed

   b. 20 in cubed

   c. 13 in cubed

   d. 100 in cubed

   e. 40 in cubed

130. Find the volume of the figure below.

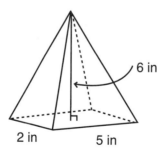

6 in

2 in   5 in

   a. 60 in cubed

   b. 40 in cubed

   c. 13 in cubed

   d. 120 in cubed

   e. 20 in cubed

Answers

86. d. Substitute the given values into the formula for volume of a cylinder. The formula uses the radius. To find the radius, divide the diameter in half. Therefore, the radius is 3 cm (6 ÷ 2 = 3); $V = \pi(3)^2 (12) = \pi(9)(12) = 108\pi$.

87. d. Substitute the given values into the formula for volume of a rectangular prism; 2.5 can be used for $2\frac{1}{2}$ to make calculations on a calculator easier; $V = (2.5)(3)(3) = 22.5$ which is equivalent to $22\frac{1}{2}$ ft cubed.

88. a. Substitute the given values into the formula for volume of a cylinder; $V = \pi(3^2)(4) = \pi(9)(4) = 36\pi$ inches cubed.

89. c. Cubic feet is the correct answer. Cubic cm, inches, and mm are too small. Cubic miles are too large.

90. a. Substitute the given values into the formula for volume of a rectangular prism; $V = (20)(8)(10) = 1{,}600$ ft cubed.

91. b. Substitute the given values into the formula for the volume of a sphere; $V = \frac{4}{3}\pi(6^3) = \frac{4}{3}\pi(216) = 288\pi$ inches cubed. Using 3.14 for π, $288 \times 3.14 = 904.32$ inches cubed.

92. e. Substitute the given values into the formula for the volume of a cone. Notice that the diameter has been given instead of the radius. Find the radius by dividing the diameter by two. Therefore, the radius of the cone is 2 cm (4 ÷ 2 = 2); $V = \frac{1}{3}\pi$ $(2^2)(10) = \frac{1}{3}(3.14)(4)(10) = 41.87$ cm cubed, when rounded to the nearest hundredth.

93. c. Substitute the given values into the formula for volume of a rectangular prism; $V = (9)(3)(4) = 108$ cm cubed.

94. d. Substitute the given values into the formula for volume of a cylinder. Notice that the diameter has been given instead of the radius. To find the radius, divide the diameter by 2. The radius is 6 feet (12 ÷ 2 = 6); $V = \pi(6^2)(4) = \pi(36)(4) = 144\pi$ ft cubed.

95. a. Substitute the given values into the formula for the volume of a sphere; $V = \frac{4}{3}\pi(10^3) = \frac{4}{3}(3.14)(1{,}000) = 4{,}186.67$ ft cubed, when rounded to the nearest hundredth.

96. **b.** Substitute the given values into the formula for the volume of a cone; $V = \frac{1}{3}\pi(4^2)(8) = \frac{1}{3}(3.14)(16)(8) = 133.97$ cm cubed, when rounded the nearest hundredth.

97. **d.** Substitute the given values into the formula for the volume of a cylinder. Note that the diameter is given instead of the radius. Find the radius by dividing the diameter by 2. The radius is 1.5 $(3 \div 2 = 1.5)$; $V = \pi(1.5^2)(6) = \pi(2.25)(6) = 13.5\pi$ m cubed.

98. **c.** Substitute the given values into the formula for the volume of a rectangular prism; $V = (5)(5)(5) = 125$ inches cubed.

99. **a.** Substitute the given values into the formula for the volume of a rectangular prism; $V = (10)(3)(14) = 420$ inches cubed.

100. **b.** Substitute the given values into the formula for the volume of a pyramid; $V = \frac{1}{3}(45)(45)(80) = 54,000$ ft cubed.

101. **e.** Substitute the given values into the formula for the volume of a cone. Note that the diameter is given instead of the radius. Find the radius by dividing the diameter by 2. The radius is 3 $(6 \div 2 = 3)$; $V = \frac{1}{3}\pi(3^2)(6) = \frac{1}{3}\pi(9)(6) = 18\pi$ inches cubed.

102. **c.** Substitute the given values into the equation for the volume of a sphere. Note that the diameter is given instead of the radius. Find the radius by dividing the diameter by 2. The radius is 15 $(30 \div 2 = 15)$; $V = \frac{4}{3}\pi(15^3) = \frac{4}{3}\pi(3,375) = 4,500\pi$ cm cubed.

103. **d.** Substitute the given values into the equation for the volume of a cone. Note that the diameter is given instead of the radius. Find the radius by dividing the diameter by 2. The radius is 5 $(10 \div 2 = 5)$; $V = \frac{1}{3}\pi(5^2)(20) = \frac{1}{3}\pi(25)(20) = 166.67\pi$ inches cubed, rounded to the nearest hundredth.

104. **a.** Substitute the given values into the equation for the volume of a sphere; $V = \frac{4}{3}\pi(7^3) = \frac{4}{3}\pi(343) = 457\pi$ inches cubed, rounded to the units place.

105. **b.** Substitute the given values into the equation for the volume of a cylinder. Note that the diameter is given instead of the radius. Find the radius by dividing the diameter by 2. The radius is 2.5 $(5 \div 2 = 2.5)$; $V = \pi(2.5^2)(7) = \pi(6.25)(7) = 43.75\pi$ m cubed.

106. b. Substitute the given values into the equation for the volume of a sphere; $V = \frac{4}{3}\pi(12^3) = \frac{4}{3}\pi(1728) = 2{,}304\pi$ cm cubed.

107. c. Substitute the given values into the formula for the volume of a rectangular prism; $V = (8)(2)(2) = 32$ m cubed.

108. e. Substitute the given values into the formula for the volume of a cylinder. Note that the diameter is given instead of the radius. Find the radius by dividing the diameter by 2. The radius is 10 feet ($20 \div 2 = 10$); $V = \pi(10^2)(35) = \pi(100)(35) = 3{,}500\pi$ ft cubed.

109. a. Substitute the given values into the formula for the volume of a pyramid; $V = \frac{1}{3}(4)(3)(5) = 20$ cm cubed.

110. c. A hemisphere is half of a sphere. To find the volume of the hemisphere, divide the volume of the sphere with the same radius by 2. The volume of a sphere with a radius of 12 m is $V = \frac{4}{3}\pi(12^3)$ $= \frac{4}{3}\pi(1{,}728) = 2{,}304\pi$ m cubed. Divide this volume by 2 to find the volume of the hemisphere; $2{,}304\pi \div 2 = 1{,}152\pi$ m cubed.

111. b. Find the volume of the entire cone (radius of 6 m and height of 10 m), then subtract the small, non-shaded cone (radius of 3 m and height of 5 m). The volume of the entire cone is $V = \frac{1}{3}\pi(6^2)(10) = 120\pi$ m cubed. The volume of the small, non-shaded cone is $V = \frac{1}{3}\pi(3^2)(5) = \frac{1}{3}\pi(9)(5) = 15\pi$ m cubed. Entire Cone – Small Cone = Shaded Portion; $120\pi - 15\pi = 105\pi$ m cubed.

112. d. Substitute the given values into the formula for volume of a rectangular prism; $V = (10)(6)(5) = 300$ inches cubed.

113. e. The length, width, and height of the cube are all 8 cm. The volume of the cube (a special type of rectangular prism) is $V = (8)(8)(8) = 512$ cm cubed.

114. c. Feet cubed is the appropriate unit of measure. Miles and km are too large. Millimeters and centimeters cubed are too small.

115. a. Substitute the given values into the formula for volume of a sphere; $V = \frac{4}{3}\pi(3^3) = \frac{4}{3}\pi(27) = 36\pi$ cm cubed.

116. d. Substitute the given values into the formula for volume of a cylinder; $V = \pi(2^2)(2) = \pi(4)(2) = 8\pi$ m cubed.

117. **b.** A can of soup is a cylinder shape. Substitute the given values into the formula for volume of a cylinder. Note that the diameter is given instead of the radius. Find the radius by dividing the diameter by 2. The radius is 1 inch ($2 \div 2 = 1$); $V = \pi(1^2)(4) = \pi(1)(4) = 4\pi$ inches cubed.

118. **c.** Substitute the given values into the equation for the volume of a pyramid; $V = \frac{1}{3}(3)(3)(8) = 24$ m cubed.

119. **e.** The globe is a sphere. Substitute the given values into the formula for the volume of a sphere. Note that the diameter is given instead of the radius. Find the radius by dividing the diameter by 2. The radius is 7 inches ($14 \div 2 = 7$); $V = \frac{4}{3}\pi(7^3) = \frac{4}{3}(3.14)(343) = 1,436.03$ inches cubed, rounded to the nearest hundredth.

120. **a.** Substitute the given values into the formula for the volume of a cone; $V = \frac{1}{3}\pi(3^2)(5) = \frac{1}{3}\pi(9)(5) = 15\pi$ cm cubed.

121. **b.** The given shape has two parts, the cylinder bottom and a hemisphere (half a sphere) top. Find both parts and add their volumes together to find the volume of the entire figure. Substitute the given values into the formula for volume of a cylinder to find the volume of the bottom part of the figure; $V = \pi(5^2)(7) = \pi(25)(7) = 549.5$ ft cubed. The radius of the hemisphere is 5 ft. Find the volume of a sphere with the same radius and divide it by 2 to find the volume of the hemisphere; $V = \frac{4}{3}\pi(5^3) = \frac{4}{3}\pi(125) = 523.33$ ft cubed. Divide the volume of the sphere by 2; $523.33 \div 2 = 261.67$ ft cubed. Add the volume of the hemisphere to the volume of the cylinder; $549.5 + 261.67 = 811.17$ ft cubed, rounded to the nearest hundredth.

122. **b.** The given shape has two parts, the cone bottom and a hemisphere (half a sphere) top. Find the volume of both parts and add their volumes together to find the volume of the entire figure. Substitute the given values into the formula for the volume of a cone to find the volume of the bottom part of the figure; $V = \frac{1}{3}\pi(4^2)(10) = \frac{1}{3}\pi(16)(10) = 167.47$ cm cubed. Find the volume of a sphere with the same radius and divide it by 2 to find the volume of the hemisphere; $V = \frac{4}{3}\pi(4^3) = \frac{4}{3}(3.14)(64) = 267.95$ cm cubed. Divide the volume of the sphere by 2; $267.95 \div 2 = 133.97$ cm cubed, rounded to the nearest hundredth. Add the volume of the hemisphere to the volume of the cone; $167.47 + 133.97 = 301.44$ cm cubed, rounded to the nearest hundredth.

123. **a.** The cement column is a cylinder. Substitute the given values into the formula for the volume of a cylinder. $V = \pi(2^2)(10) = \pi(4)(10) = 40\pi$ ft cubed.

124. **d.** Substitute the given values into the formula for the volume of a pyramid; $V = \frac{1}{3}(3)(4)(10) = 40$ m cubed.

125. **b.** Substitute the given values into the formula for volume of a rectangular prism; $V = (6)(3)(2) = 36$ inches cubed.

126. **c.** Substitute the given values into the formula for the volume of a rectangular prism; $V = (18)(24)(15) = 6{,}480$ inches cubed.

127. **a.** Substitute the given values into the formula for the volume of a sphere; $V = \frac{4}{3}\pi(9^3) = \frac{4}{3}\pi(729) = 972\pi$ m cubed.

128. **d.** Substitute the given values into the formula for the volume of a cone. Note that that diameter is given instead of the radius. Find the radius by dividing the diameter by 2. The radius is 7 cm ($14 \div 2 = 7$); $V = \frac{1}{3}\pi(7^2)(12) = \frac{1}{3}\pi(49)(12) = 196\pi$ cm cubed.

129. **a.** The box is a rectangular prism. Substitute the given values into the formula for the volume of a rectangular prism; $V = (5)(4)(4) = 80$ inches cubed.

130. **e.** Substitute the given values into the formula for the volume of a pyramid; $V = \frac{1}{3}(5)(2)(6) = 20$ inches cubed.

4

Measurement— Angles and Arc

Angles are one of the basic geometric figures. Angles are classified and named according to their degree size. A protractor measures angles. There are various angle pairs that have special relationships. Look over the formulas that follow to review these aspects of angles. You will be expected to know these relationships and vocabulary. An arc is a part of a circle, and, like angles, is measured in degrees. Arcs are not measured by a protractor, but are calculated from information about related angles in the interior or exterior of the circle. In addition to the formula sheet, the answer explanations give detailed methods of solving a variety of problems. By working through the problems in this chapter, you will effectively review the concepts that deal with angles and arc.

Formulas

Types of Angles

Acute—measures between 0° and 90°
Obtuse—measures between 90° and 180°
Right—measures exactly 90°
Straight—measures exactly 180°
Reflex—measures more than 180°

The sum of the measures of two complementary angles is 90°.

The sum of the measures of two supplementary angles is 180°.

Vertical angles formed by two intersecting lines are congruent.

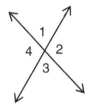

$$\angle 1 \cong \angle 3, \angle 2 \cong \angle 4$$

There are special angle pairs formed by two parallel lines cut by a transversal.

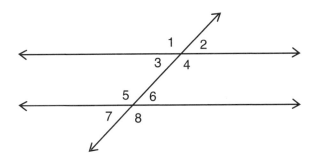

Alternate interior angles are congruent; $\angle 3 \cong \angle 6$, $\angle 4 \cong \angle 5$.
Alternate exterior angles are congruent; $\angle 1 \cong \angle 8$, $\angle 2 \cong \angle 7$.
Corresponding angles are congruent; $\angle 1 \cong \angle 5$, $\angle 2 \cong \angle 6$, $\angle 3 \cong \angle 7$, $\angle 4 \cong \angle 8$.
Vertical angles are congruent; $\angle 1 \cong \angle 4$, $\angle 2 \cong \angle 3$, $\angle 5 \cong \angle 8$, $\angle 6 \cong \angle 7$.

Angles of polygons

Sum of the interior angles is $180(n - 2)$ where n is the number of sides.
Sum of the exterior angles is always $360°$.

Arc measure

Measure of a central angle is equal to the measure of the intercepted arc.

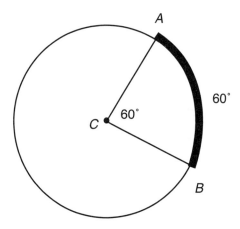

Measure of an inscribed angle is one-half the measure of the intercepted arc.

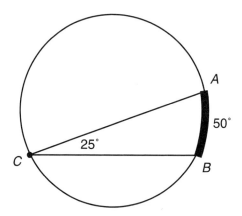

Measure of an angle formed by two intersecting chords in a circle is equal to half the sum of the intercepted arcs.

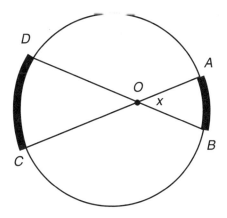

$$m\angle ACB = \tfrac{1}{2}(m\,\overarc{AB} - m\,\overarc{CD})$$

Measure of an angle in the exterior of a circle formed by two secants, two tangents, or a tangent and a secant is equal to half the difference of the intercepted arcs.

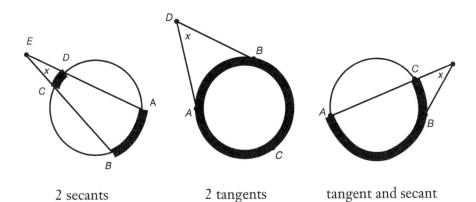

2 secants   2 tangents   tangent and secant

$$m\angle DEC = \tfrac{1}{2}(m\,\overarc{AB} - m\,\overarc{CD}) \quad m\angle BDA = \tfrac{1}{2}(m\,\overarc{ACB} - m\,\overarc{AB}) \quad m\angle BDC = \tfrac{1}{2}(m\,\overarc{AB} - m\,\overarc{BC})$$

131. Which of the following units is used to measure angles?
   a. meters
   b. inches
   c. square inches
   d. degrees
   e. pixels

132. An angle that measures 65° can be classified as a(n)
   a. right angle.
   b. acute angle.
   c. obtuse angle.
   d. straight angle.
   e. reflex angle.

133. The angle in the figure below can be best described as a(n)

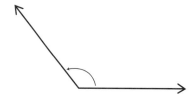

   a. acute angle.
   b. straight angle.
   c. obtuse angle.
   d. right angle.
   e. reflex angle.

134. A right angle is an angle that measures
   a. less than 90°.
   b. more than 90°.
   c. exactly 90°.
   d. exactly 180°.
   e. more than 180°.

135. If an angle measures between 180° and 360°, it can be classified as a(n)

    a. right angle.

    b. straight angle.

    c. obtuse angle.

    d. reflex angle.

    e. acute angle.

136. What is the sum of the measure of an angle and its complement?

    a. 30°

    b. 45°

    c. 60°

    d. 90°

    e. 180°

137. What is the measure of $\angle ABD$ in the following diagram?

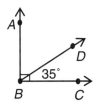

    a. 35°

    b. 45°

    c. 55°

    d. 90°

    e. 145°

138. When measuring the number of degrees in an angle, which measurement tool should you use?

    a. ruler

    b. protractor

    c. measuring tape

    d. thermometer

    e. scale

139. If the angles in the following diagram are complementary, what is the measure of ∠B?

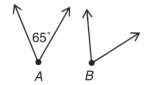

a. 15°

b. 25°

c. 35°

d. 115°

e. 135°

140. What is the measure of the complement of a 70° angle?

a. 30°

b. 20°

c. 10°

d. 110°

e. none of these

141. If the angles in the diagram are supplementary, what is the measure of ∠D?

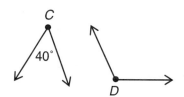

a. 50°

b. 60°

c. 100°

d. 140°

e. 160°

142. What is the measure of the supplement of a 123° angle?

    a. 90°

    b. 77°

    c. 67°

    d. 57°

    e. 180°

143. What is the measure of ∠*ABD* in the following diagram?

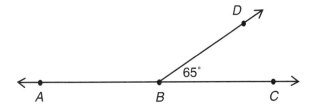

    a. 90°

    b. 25°

    c. 125°

    d. 135°

    e. 115°

144. What is the measure of ∠1 in the following diagram?

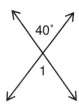

    a. 10°

    b. 40°

    c. 50°

    d. 140°

    e. 110°

145. What is the measure of ∠1 in the following diagram?

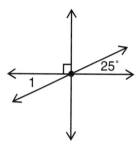

    a. 25°

    b. 35°

    c. 65°

    d. 155°

    e. 115°

146. What is the measure of ∠1 in the following diagram?

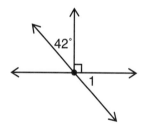

    a. 42°

    b. 48°

    c. 52°

    d. 58°

    e. 138°

147. If ray $\overrightarrow{BD}$ bisects $\angle ABC$ and the measure of $\angle ABD$ is 30°, what is the measure of $\angle ABC$?

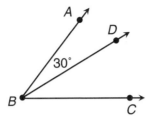

a. 15°
b. 30°
c. 45°
d. 60°
e. 120°

148. Ray $\overrightarrow{XZ}$ bisects $\angle WXY$, the measure of $\angle WXZ = 2x$ and the measure of $\angle YXZ = x + 75$. What is the measure of $\angle WXZ$?

a. 15°
b. 37.5°
c. 45°
d. 75°
e. 150°

Use the following diagram of two parallel lines cut by a transversal for questions 149–151.

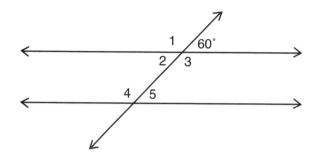

149. What is the measure of ∠1?

    a. 60°

    b. 90°

    c. 130°

    d. 120°

    e. 140°

150. Which of the angles shown in the diagram are congruent to ∠4?

    a. ∠1

    b. ∠3

    c. both ∠1 and ∠3

    d. both ∠2 and ∠5

    e. ∠2

151. Which of the following angles are supplementary to ∠5?

    a. ∠2

    b. ∠3

    c. Both ∠3 and ∠4

    d. Both ∠1 and ∠2

    e. ∠1, ∠3, and ∠4

152. In the following diagram of two parallel lines cut by a transversal, which angle is an alternate interior angle with respect to ∠2?

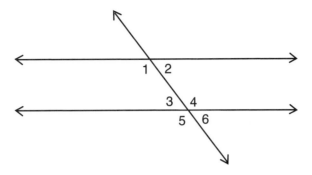

a. ∠1
b. ∠3
c. ∠4
d. ∠5
e. ∠6

153. Given the following diagram of two parallel lines cut by a transversal, what is the value of *x*?

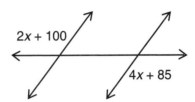

a. 0
b. 5
c. 7.5
d. 15
e. 2.5

154. In △ABC, the measure of ∠A is 32° and the measure of ∠B is 78°. What is the measure of ∠C?

    a. 12°

    b. 20°

    c. 70°

    d. 78°

    e. 250°

155. If the vertex angle of isosceles triangle XYZ measures 50°, what is the measure of base angle ∠Z?

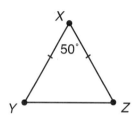

    a. 50°

    b. 65°

    c. 75°

    d. 130°

    e. 32.5°

156. In quadrilateral $MNOP$, the measure of ∠M is 110°, the measure of ∠N is 90°, and the measure of ∠O is 95°. What is the measure of ∠P?

    a. 65°

    b. 75°

    c. 85°

    d. 105°

    e. 165°

157. What is the sum of the interior angles of a regular pentagon?

    a. 90°

    b. 180°

    c. 108°

    d. 360°

    e. 540°

158. What is the degree measure of one interior angle of a regular hexagon?

    a. 72°

    b. 90°

    c. 180°

    d. 720°

    e. 120°

159. What is the sum of the degree measure of the exterior angles of a regular octagon?

    a. 45°

    b. 72°

    c. 360°

    d. 90°

    e. 108°

160. What is the measure of an exterior angle of an equilateral triangle?

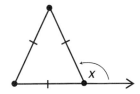

    a. 60°

    b. 90°

    c. 180°

    d. 120°

    e. 30°

161. The sum of the measures of the arcs that form a circle is always

    a. 90°.

    b. 180°.

    c. 270°.

    d. 360°

    e. 540°

162. The measure of an arc intercepted by a central angle of a circle
   a. is equal to the central angle.
   b. is one-half of the central angle.
   c. is twice the degree measure of the central angle.
   d. is always 90°.
   e. cannot be determined.

163. $\angle ACB$ is a central angle. What is the measure of $\overset{\frown}{AB}$?

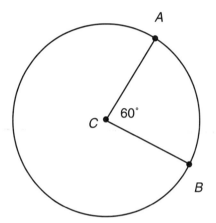

   a. 15°
   b. 30°
   c. 60°
   d. 90°
   e. 120°

164. Central angle $\angle XYZ$ intercepts minor $\overarc{XZ}$. If the measure of $\overarc{XZ}$ is 110°, what is the measure of $\angle XYZ$?

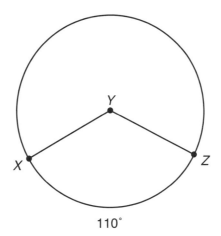

110˚

a. 70°

b. 55°

c. 220°

d. 250°

e. 110°

165. The measure of central angle ∠*AOB* is 90°. What is the measure of major $\overset{\frown}{ACB}$?

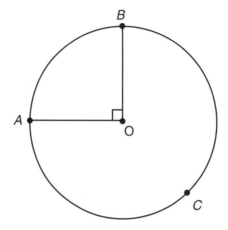

    a. 45°

    b. 90°

    c. 180°

    d. 270°

    e. 360°

166. The measure of an inscribed angle is _____ the measure of the arc it intercepts.

    a. equal to

    b. twice

    c. one-half

    d. four-times

    e. not related to

167. In the following diagram, inscribed angle $\angle ABC$ intercepts $\overset{\frown}{AC}$. If the measure of $\overset{\frown}{AC}$ is 30°, what is the measure of $\angle ABC$?

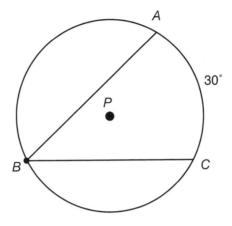

    a. 15°

    b. 30°

    c. 45°

    d. 60°

    e. none of these

168. Within a circle, inscribed $\angle XYZ$ has a measure of 25° and intercepts minor $\overset{\frown}{XZ}$. What is the measure of $\overset{\frown}{XZ}$?

    a. 12.5°

    b. 25°

    c. 50°

    d. 75°

    e. 100°

169. In the following figure, $\overline{CD}$ is a diameter of the circle and $\overline{BA}$ is a radius. If the measure of $\overset{\frown}{AC}$ is 125°, what is the measure of $\overset{\frown}{AD}$?

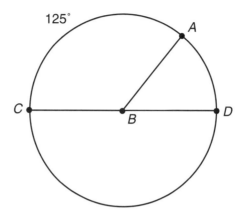

125°

    a. 125°

    b. 62.5°

    c. 27.5°

    d. 55°

    e. 180°

170. In the following figure, the ratio of the arcs $\overset{\frown}{WX} : \overset{\frown}{XY} : \overset{\frown}{YZ} : \overset{\frown}{ZW}$ is 2 : 2 : 1 : 1. What is the measure of $\overset{\frown}{XY}$?

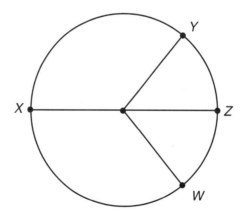

    a. 60°

    b. 30°

    c. 120°

    d. 90°

    e. 180°

171. Use the following circle to find the measure of ∠*YPZ*.

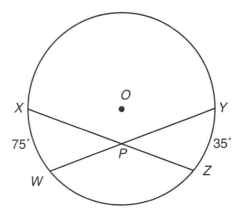

    a. 35°

    b. 45°

    c. 55°

    d. 65°

    e. 75°

172. Given that $\overline{AC}$ is a diameter of the circle below, what is the measure of $\overset{\frown}{AB}$?

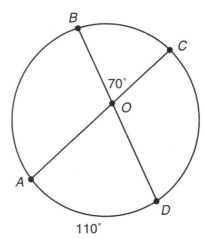

    a. 60°

    b. 140°

    c. 70°

    d. 110°

    e. 150°

173. Given the following figure with two tangents drawn to the circle, what is the measure of ∠*ACB*?

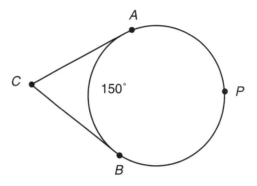

   a. 30°

   b. 15°

   c. 60°

   d. 120°

   e. 75°

174. Given the following figure with two secants drawn to the circle, what is the measure of ∠*AED*?

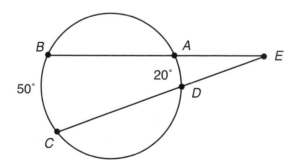

   a. 60°

   b. 30°

   c. 120°

   d. 15°

   e. 35°

175. Given the following figure with one tangent and one secant drawn to the circle, what is the measure of ∠*ADB*?

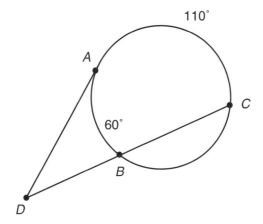

a. 50°

b. 85°

c. 60°

d. 110°

e. 25°

Answers

131. **d.** Degrees are used to measure angles. Meters, inches, and pixels are all units of length and square inches are a measure of area.

132. **b.** An angle that measures between 0° and 90° is called an acute angle.

133. **c.** Since this angle is greater than 90° but less than 180° it is an obtuse angle.

134. **c.** A right angle is an angle that measures exactly 90°.

135. **d.** An angle that measures more than 180° is called a reflex angle.

136. **d.** If two angles are complementary, the sum of their angle measures is 90°. Therefore, the measure of an angle and its complement is 90°.

137. **c.** Since the two angles together form a right angle and a right angle always measures 90°, subtract to find the measure of $\angle ABD$; 90 − 35 = 55°.

138. **b.** A protractor is used to measure the degrees of an angle. Both a ruler and measuring tape are used to measure length. A thermometer is used to measure temperature and a scale measures the weight of an object.

139. **b.** If two angles are complementary, the sum of their angle measures is 90°. Subtract 90 − 65 to get the measure of $\angle ABC$, which is 25°.

140. **b.** If two angles are complementary, the sum of their angle measures is 90°. Subtract 90 − 70 to get the measure of the complement, which is 20°.

141. **d.** If two angles are supplementary, the sum of their angle measures is 180°. Subtract 180 − 40 to get the measure of $\angle D$, which is 140°.

142. **d.** If two angles are supplementary, the sum of their angle measures is 180°. Subtract 180 − 123 to get the measure of the supplement, which is 57°.

143. **e.** The two angles in the diagram together form a straight angle, which is equal to 180°. Therefore, the two angles are supplementary. Subtract 180 – 65 to find the measure of ∠ABD, which is 115°.

144. **b.** The angle labeled 40° and ∠1 are vertical angles; therefore their measures are equal. The measure of ∠1 is 40°.

145. **a.** The same two intersecting lines form the angle labeled 25° and ∠1 in the diagram. Therefore, they are vertical angles and have the same measure. The measure of ∠1 is 25°.

146. **b.** The angle labeled 42°, the angle marked a right angle, and ∠1 together form a straight angle. To find the measure of ∠1, subtract 180 – (42 + 90); 180 – 132 = 48°.

147. **d.** A ray that bisects an angle divides the angle into two equal parts. If the measure of ∠ABD is 30°, then the measure of the entire angle ∠ABC is twice that, or 60°.

148. **e.** Since ray $\overrightarrow{XZ}$ bisects ∠WXY, the measure of the two resulting angles ∠WXZ and ∠YXZ are equal. Therefore, set the two expressions for the angles equal to each other and solve for x. Write the equation: $2x = x + 75$. Subtract x from both sides of the equation: $2x - x = x - x + 75$. So, $x = 75$. The value of ∠WXZ is $2x = 2(75) = 150°$.

149. **d.** The angle labeled 60° and ∠1 form a straight angle, and therefore the sum of their angle measures is 180°. Subtract 180 – 60 to find the measure of ∠1, which is 120°.

150. **c.** ∠1 is congruent to ∠4 because they are corresponding angles; ∠3 is congruent to ∠4 because they are alternate interior angles.

151. **e.** Since ∠4 forms a straight angle with ∠5, they are supplementary angles. Any angle congruent to ∠4 will also be supplementary to ∠5; ∠1 and ∠4 are corresponding angles, so they are congruent. ∠3 and ∠4 are alternate interior angles, so they are also congruent.

152. **b.** Alternate interior angles are two angles located between two parallel lines on opposite sides of the transversal; ∠2 and ∠3 are alternate interior angles.

153. **c.** The angles labeled in the diagram are alternate exterior angles so their measures are equal. Write an equation setting the expressions for each angle equal to each other and solve for x; $2x + 100 = 4x + 85$. Subtract $2x$ from each side of the equal sign; $2x - 2x + 100 = 4x - 2x + 85$. Simplify; $100 = 2x + 85$. Subtract 85 from each side of the equation and combine like terms; $100 - 85 = 2x + 85 - 85$; $15 = 2x$. Divide each side of the equation by 2 to get $x = 7.5$.

154. **c.** The sum of the measures of the interior angles of a triangle is 180. Subtract the sum of the two known angles from 180 to find the missing angle; $180 - (78 + 32) = 180 - 110 = 70$.

155. **b.** The sum of the measures of the interior angles of a triangle is 180, so subtract the measure of the vertex angle from 180 to find the remaining degrees of the triangle; $180 - 50 = 130$. Since the base angles of an isosceles triangle are congruent, divide 130 by 2 to find the measure of $\angle Z$, one of the base angles; $130 \div 2 = 65°$.

156. **a.** The sum of the measures of the interior angles of any polygon can be found using the formula $180(n - 2)$ where n is the number of sides of the polygon. Therefore, the number of degrees in the interior angles of a quadrilateral are $180(4 - 2) = 180(2) = 360$. Subtract the sum of the known angles of the quadrilateral from 360 to find the measure of $\angle P$; $360 - (110 + 90 + 95) = 360 - 295 = 65°$.

157. **e.** The sum of the measures of the interior angles of any polygon can be found using the formula $180(n - 2)$ where n is the number of sides of the polygon. Therefore, the number of degrees in the interior angles of a pentagon are $180(5 - 2) = 180(3) = 540$.

158. **e.** First find the measure of the sum of the interior angles of a hexagon. The sum of the measures of the interior angles of any polygon can be found using the formula $180(n - 2)$ where n is the number of sides of the polygon. Therefore, the number of degrees in the interior angles of a hexagon are $180(6 - 2) = 180(4) = 720$. Since a regular hexagon has six congruent sides and six congruent angles, divide 720 by 6 to find the measure of one interior angle; $720 \div 6 = 120°$.

159. **c.** The sum of the degree measure of the exterior angles of any polygon is $360°$.

160. **d.** An equilateral triangle has three congruent sides and three congruent angles. The interior angles are congruent, so therefore the exterior angles are also congruent. Since the sum of the degree measure of the exterior angles of any polygon is 360, divide 360 by 3 to find the measure of one exterior angle of an equilateral triangle; $360 \div 3 = 120°$.

161. **d.** The sum of the measures of the arcs that form a full circle is always 360°.

162. **a.** A central angle of a circle is an angle whose vertex is the center of the circle and whose sides are radii of the circle. The measure of the central angle is always equal to the arc of the circle that it intercepts.

163. **c.** A central angle of a circle is an angle whose vertex is the center of the circle and whose sides are radii of the circle. Since $\angle ACB$ is a central angle, the measure of the intercepted $\overset{\frown}{AB}$ is equal to the measure of the angle. Therefore, the measure of $\overset{\frown}{AB}$ is 60°.

164. **e.** A central angle of a circle is an angle whose vertex is the center of the circle and whose sides are radii of the circle. Since $\angle XYZ$ is a central angle, the measure of the intercepted $\overset{\frown}{XZ}$ is equal to the measure of the angle. Therefore, the measure of $\angle XYZ$ is 110°.

165. **d.** A central angle of a circle is an angle whose vertex is the center of the circle and whose sides are radii of the circle. Since (AOB is a central angle, the measure of the intercepted $\overset{\frown}{AB}$ is equal to the measure of the angle. So, the measure of $\overset{\frown}{AB}$ is 90°. To find the measure of major $\overset{\frown}{ACB}$, subtract 90 from the total degrees in the arcs of the circle, which is 360; $360 - 90 = 270°$.

166. **c.** An inscribed angle is an angle with its vertex on the circle and whose sides are chords of the circle. The measure of an inscribed angle is equal to one-half the measure of the arc it intercepts.

167. **a.** An inscribed angle is an angle with its vertex on the circle and whose sides are chords of the circle. The measure of an inscribed angle is equal to one-half the measure of the arc it intercepts. Since the measure of $\overset{\frown}{AC}$ is 30°, the measure of $\angle ABC$ is $\frac{1}{2}$ of 30, or 15°.

168. **c.** An inscribed angle is an angle with its vertex on the circle and whose sides are chords of the circle. The measure of an inscribed angle is equal to one-half the measure of the arc it intercepts. Since the measure of $\angle XYZ$ is 25°, the measure of $\overset{\frown}{XZ}$ is twice the measure of the angle, or 50°.

169. **d.** Since $\overline{CD}$ is a diameter of the circle, it divides the circle into two congruent parts. Thus, the measure of $\overset{\frown}{CAD}$ is one-half of 360, or 180°. Since the measure of $\overset{\frown}{AC}$ is 125°, subtract 125 from 180 to find the measure of $\overset{\frown}{AD}$; 180 − 125 = 55°.

170. **c.** The sum of all of the arcs of a circle equals 360°. Use this fact and multiply each of the values in the ratio by x to write an equation; $2x + 2x + 1x + 1x = 360$. Combine like terms; $6x = 360$. Divide each side of the equation by 6; $x = 60$. Since the value of $\overset{\frown}{XY}$ is $2x$, $2(60) = 120°$.

171. **c.** The measure of an angle formed by two intersecting chords in a circle is equal to half the sum of the intercepted arcs. The intercepted arcs $\overset{\frown}{XW}$ and $\overset{\frown}{YZ}$ measure 75° and 35°, respectively. Half of the sum of the intercepted arcs can be expressed as $\frac{1}{2}(75 + 35) = \frac{1}{2}(110) = 55°$. Therefore, the measure of $\angle YPZ$ is 55°.

172. **e.** In this problem, use the measures of the known arcs and angles first to help find the measure of $\overset{\frown}{AB}$. The measure of an angle formed by two intersecting chords in a circle is equal to half the sum of the intercepted arcs. Since the measure of $\angle BOC$ is 70° and the intercepted $\overset{\frown}{AD}$ is 110°, then $70 = \frac{1}{2}(110 + \text{measure of } \overset{\frown}{BC})$. Multiply each side of the equal sign by 2 to get $140 = 110 + (\text{measure of } \overset{\frown}{BC})$. Subtract 110 from each side to get $30 = \text{measure}$ of $\overset{\frown}{BC}$. Since $\overline{AC}$ is a diameter of the circle, the measure of $\overset{\frown}{ABC}$ = 180°. Therefore, the measure of $\overset{\frown}{AB}$ is equal to 180 − 30 = 150°.

173. **a.** The measure of an angle in the exterior of a circle formed by two tangents is equal to half the difference of the intercepted arcs. The two intercepted arcs are $\overset{\frown}{AB}$, which is 150°, and $\overset{\frown}{APB}$, which is 360 − 150 = 210°. Find half of the difference of the two arcs; $\frac{1}{2}(210 - 150) = \frac{1}{2}(60) = 30°$. The measure of $\angle ACB$ is 30°.

174. **d.** The measure of an angle in the exterior of a circle formed by two secants is equal to half the difference of the intercepted arcs. The two intercepted arcs are $\overset{\frown}{AD}$, which is 20°, and $\overset{\frown}{BC}$, which is 50°. Find half of the difference of the two arcs; $\frac{1}{2}(50 - 20) = \frac{1}{2}(30) = 15°$. The measure of $\angle AED$ is 15°.

175. **e.** The measure of an angle in the exterior of a circle formed by a tangent and a secant is equal to half the difference of the intercepted arcs. The two intercepted arcs are $\overset{\frown}{AB}$, which is 60°, and $\overset{\frown}{AC}$, which is 110°. Find half of the difference of the two arcs; $\frac{1}{2}(110 - 60) = \frac{1}{2}(50) = 25°$. The measure of $\angle ADB$ is 25°.

5

Conversion—Time

This chapter covers time conversions. Time is measured in many different forms. The history of a country can be measured in centuries; a human lifetime can be measured in decades or years. In planning a vacation, days and weeks are important; in planning a day, hours, minutes, and even seconds become critical. After solving a problem, be sure to use common sense and ask yourself if the answer you arrive at is reasonable. If a question asks how many seconds are in two hours, for example, you have the common sense to know that the answer will be a number much larger than two. Many of the problems will require you to make more than one conversion to arrive at a solution, as in the example above; hours would be first converted to minutes and then minutes to seconds. Try these questions and read over the answer explanations. By the "time" you finish this chapter you will have mastered this type of conversion.

176. It took Kaitlyn two hours to finish her homework. How many minutes did it take her to finish her homework?
   a. 90 minutes
   b. 60 minutes
   c. 100 minutes
   d. 120 minutes
   e. 150 minutes

177. Michael takes four minutes to shave each morning. How many seconds does Michael spend shaving each morning?
   a. 240 seconds
   b. 64 seconds
   c. 200 seconds
   d. 120 seconds
   e. 400 seconds

178. Bill worked for a steel manufacturer for three decades. How many years did Bill work for the steel manufacturer?
   a. 15 years
   b. 60 years
   c. 30 years
   d. 6 years
   e. 12 years

179. How many minutes are there in 12 hours?
   a. 24 minutes
   b. 1,440 minutes
   c. 36 minutes
   d. 1,200 minutes
   e. 720 minutes

180. Millie's ancestors first arrived in the United States exactly two centuries ago. How many years ago did Millie's ancestors arrive in the United States?
   a. 200 years
   b. 50 years
   c. 20 years
   d. 100 years
   e. 30 years

181. How many days are there in a leap year?
   a. 375 days
   b. 366 days
   c. 364 days
   d. 325 days
   e. 397 days

182. Mrs. Dias gave her math class exactly $1\frac{1}{2}$ hours to complete their final exam. How many minutes did the class have to finish the exam?
   a. 30 minutes
   b. 90 minutes
   c. 150 minutes
   d. 60 minutes
   e. 120 minutes

183. Rita is eight decades old. How many years old is Rita?
   a. 40 years old
   b. 16 years old
   c. 64 years old
   d. 48 years old
   e. 80 years old

184. Lucy spent 240 minutes working on her art project. How many hours did she spend on the project?
   a. 14,400 hours
   b. 24 hours
   c. 4 hours
   d. 2 hours
   e. 6 hours

185. How many days comprise two years (assume that neither year is a leap year)?
   a. 200 days
   b. 400 days
   c. 624 days
   d. 730 days
   e. 750 days

186. Jessica is 17 years old today. How many months old is she?

    a. 204 months

    b. 255 months

    c. 300 months

    d. 221 months

    e. 272 months

187. Lou just completed 48 months of service in the Army. How many years was Lou in the Army?

    a. 12 years

    b. 4 years

    c. 6 years

    d. 10 years

    e. 15 years

188. How many years are there in five centuries?

    a. 35 years

    b. 125 years

    c. 250 years

    d. 400 years

    e. 500 years

189. Popcorn is microwaved for $3\frac{1}{2}$ minutes. How many seconds is the popcorn microwaved?

    a. 180 seconds

    b. 175 seconds

    c. 80 seconds

    d. 60 seconds

    e. 210 seconds

190. How many hours are there in one week?

    a. 168 hours

    b. 120 hours

    c. 84 hours

    d. 144 hours

    e. 96 hours

191. Marty has lived in his house for 156 weeks. How many years has he lived in his house?
a. 2 years
b. 5 years
c. 3 years
d. 11 years
e. 9 years

192. How many hours are in five days?
a. 60 hours
b. 100 hours
c. 120 hours
d. 240 hours
e. 600 hours

193. 2,520 seconds is equivalent to how many minutes?
a. 84 minutes
b. 42 minutes
c. 151,200 minutes
d. 126 minutes
e. 252 minutes

194. A scientist has spent the past two decades studying a specific type of frog. How many years has the scientist spent studying this frog?
a. 10 years
b. 12 years
c. 15 years
d. 20 years
e. 30 years

195. Josh slept 540 minutes last night. How many hours did he sleep?
a. 9 hours
b. 8 hours
c. 7 hours
d. 6 hours
e. 5 hours

196. How many seconds are in two hours?
   a. 120 seconds
   b. 240 seconds
   c. 2,400 seconds
   d. 72,000 seconds
   e. 7,200 seconds

197. Betsy and Tim have been married 30 years. How many decades have they been married?
   a. 300 decades
   b. 3 decades
   c. 10 decades
   d. 6 decades
   e. 600 decades

198. Jacob is 7,200 minutes old. How many days old is Jacob?
   a. 120 days
   b. 5 days
   c. 24 days
   d. 36 days
   e. 2 days

199. How many decades are in five centuries?
   a. 500 decades
   b. 100 decades
   c. 40 decades
   d. 50 decades
   e. 25 decades

200. Today is Rosa's seventh birthday. How many months old is Rosa?
   a. 84 months
   b. 70 months
   c. 74 months
   d. 90 months
   e. 104 months

201. Kyra's biology class is 55 minutes long. How many seconds long is the class?
  a. 3,000 seconds
  b. 4,400 seconds
  c. 5,400 seconds
  d. 8,600 seconds
  e. 3,300 seconds

202. How many months are there in six decades?
  a. 60 months
  b. 720 months
  c. 48 months
  d. 660 months
  e. 960 months

203. Two hundred and forty months have passed since Guy started his current job. How many decades has Guy worked at his current job?
  a. 20 decades
  b. 4 decades
  c. 2 decades
  d. 10 decades
  e. 12 decades

204. Peter is going on vacation in exactly five weeks. How many days until Peter goes on vacation?
  a. 45 days
  b. 30 days
  c. 35 days
  d. 40 days
  e. 50 days

205. How many seconds are there in nine hours?
  a. 32,400 seconds
  b. 540 seconds
  c. 60 seconds
  d. 44,000 seconds
  e. 960 seconds

206. Wendy took out a car loan for 60 months. How many years is the loan?
   a. 6 years
   b. 5 years
   c. 4 years
   d. 3 years
   e. 2 years

207. Lois spent a total of four hours working on her science project. How many minutes did Lois spend on the project?
   a. 48 minutes
   b. 96 minutes
   c. 108 minutes
   d. 220 minutes
   e. 240 minutes

208. Two centuries have passed since Joan's house was built. How many months have passed since Joan's house was built?
   a. 200 months
   b. 1,200 months
   c. 2,400 months
   d. 8,000 months
   e. 10,400 months

Answers

176. **d.** Multiply the number of minutes in an hour by the given number of hours. There are 60 minutes in an hour. Therefore, there are 120 minutes in 2 hours; 2 hours $\times$ 60 minutes = 120 minutes.

177. **a.** Multiply the number of seconds in a minute by the given number of minutes. There are 60 seconds in one minute. There are 240 seconds in 4 minutes; 4 minutes $\times$ 60 seconds = 240 seconds.

178. **c.** Multiply the number of years in a decade by the given number of decades. There are 10 years in a decade. Three decades is 30 years; 3 decades $\times$ 10 years = 30 years.

179. **e.** Multiply the number of minutes in an hour by the given number of hours. There are 60 minutes in each hour. Therefore, there are 720 minutes in 12 hours; 12 hours $\times$ 60 minutes = 720 minutes.

180. **a.** Multiply the number of years in a century by the given number of centuries. There are 100 years in a century. Two centuries is 200 years; 2 centuries $\times$ 100 years = 200 years.

181. **b.** A leap year happens every four years and it has one more day than a regular year, February 29. A regular year has 365 days, so a leap year has 366 days.

182. **b.** Multiply the number of minutes in an hour by the given number of hours. An hour has 60 minutes; $1\frac{1}{2}$ hours is equivalent to 90 minutes; $1\frac{1}{2}$ hours $\times$ 60 minutes = 90 minutes. You can also think of it as one hour (60 minutes) plus another $\frac{1}{2}$ hour (30 minutes) which is a total of 90 minutes.

183. **e.** Multiply the number of years in a decade by the given number of decades. A decade is ten years. Eight decades is therefore 80 years; 8 decades $\times$ 10 years = 80 years.

184. **c.** Divide the total number of minutes by the number of minutes in an hour. There are 60 minutes in an hour. There are 4 hours in 240 minutes; 240 minutes $\div$ 60 minutes = 4 hours.

185. **d.** Multiply the number of days in a year by the given number of years. A year is 365 days. Two years is 730 days; 365 days $\times$ 2 years = 730 days.

186. **a.** Multiply the number of months in a year by the given number of years. There are 12 months in a year. There are 204 months in 17 years; 17 years $\times$ 12 months = 204 months.

187. **b.** Divide the total number of months by the number of months in a year. Each year is 12 months. There are 4 years in 48 months; 48 months $\div$ 12 months = 4 years.

188. **e.** Multiply the number of years in a century by the given number of centuries. There are 100 years in a century. Therefore, there are 500 years in 5 centuries; 5 centuries $\times$ 100 years = 500 years.

189. **e.** Multiply the number of seconds in a minute by the given number of minutes. There are 60 seconds in a minute. There are 210 seconds in $3\frac{1}{2}$ minutes; $3\frac{1}{2}$ minutes $\times$ 60 seconds = 210 seconds. You can also think of it as 180 seconds in 3 minutes (3 minutes $\times$ 60 seconds = 180 seconds) plus 30 seconds in $\frac{1}{2}$ a minute (1\2 minute $\times$ 60 seconds = 30 seconds), therefore, there are 210 seconds in $3\frac{1}{2}$ minutes; 180 seconds + 30 seconds = 210 seconds.

190. **a.** Multiply the number of days in one week by the number of hours in each day. First, there are seven days in a week. Next, there are 24 hours in each of those days. Therefore, there are 168 hours in a week; 7 days $\times$ 24 hours = 168 hours.

191. **c.** Divide the total number of weeks by the number of weeks in a year. Each year is made up of 52 weeks. There are 3 years in 156 weeks; 156 weeks $\div$ 52 weeks = 3 years.

192. **c.** Multiply the number of hours in a day by the given number of days. There are 24 hours in each day. There are 120 hours in 5 days; 5 days $\times$ 24 hours = 120 hours.

193. **b.** Divide the total number of seconds by the number of seconds in a minute. There are 60 seconds in a minute; 2,520 seconds is 42 minutes; 2,520 seconds $\div$ 60 seconds = 42 minutes.

194. d. Multiply the number of years in a decade by the given number of decades. A decade is ten years; 2 decades is 20 years; 10 years × 2 decades = 20 years.

195. a. Divide the total number of minutes by the number of minutes in an hour. An hour is 60 minutes; 540 minutes is 9 hours; 540 minutes ÷ 60 minutes = 9 hours.

196. e. First, find the number of minutes in 2 hours. There are 60 minutes in one hour, so there are 120 minutes in 2 hours. Next, find the number of seconds in those 120 minutes. There are 60 seconds in a minute. Therefore, there are 7,200 seconds in 120 minutes; 120 minutes × 60 seconds = 7,200 seconds.

197. b. Divide the total number of years by the number years in a decade. There are ten years in a decade; 30 years is 3 decades; 30 years ÷ 10 years = 3 decades.

198. b. You must change minutes to hours, then hours to days. First, there are 60 minutes in an hour; 7,200 minutes is 120 hours (7,200 minutes ÷ 60 minutes = 120 hours). Next, there are 24 hours in a day; 120 hours is 5 days; 120 hours ÷ 24 hours = 5 days.

199. d. First, determine how many decades are in one century. There are 100 years in a century, and ten years in a decade. There are ten decades in a century (100 years ÷ 10 years = 10 decades). Therefore, there are fifty decades in five centuries; 5 centuries × 10 decades = 50 decades.

200. a. Multiply the number of months in a year by the given number of years. There are 12 months in a year. There are 84 months in 7 years; 7 years × 12 months = 84 months.

201. e. Multiply the number of seconds in a minute by the given number of minutes. A minute is sixty seconds; 55 minutes is 3,300 seconds; 55 minutes × 60 seconds = 3,300 seconds.

202. b. First, determine how many years are in six decades. Since there are ten years in a decade, there are sixty years in six decades (6 decades × 10 years = 60 years). Next, multiply this number of years by the number of months in one year. Since there are 12 months in one year, there are 720 months in 60 years (60 years × 12 months = 720 months).

203. c. You must convert the months to years and then, years to decades. First, divide the total number of months by the number of months in one year. There are 12 months in a year. Therefore, there are 20 years in 240 months (240 months ÷ 12 months = 20 years). Now, divide this total number of years by the number of years in one decade. There are ten years in a decade. There are two decades in 20 years; 20 years ÷ 10 years = 2 decades.

204. c. Multiply the number of days in a week by the given number of weeks. There are seven days in a week. There are 35 days in five weeks; 5 weeks × 7 days = 35 days.

205. a. First convert hours to minutes, and then change minutes to seconds. There are sixty minutes in an hour. Therefore, there are 540 minutes in nine hours (9 hours × 60 minutes = 540 minutes). Next, there are 60 seconds in a minute. Therefore, there are 32,400 seconds in 540 minutes; 540 minutes × 60 seconds = 32,400 seconds.

206. b. Divide the total number of months by the number of months in a year. There are 12 months in a year. Sixty months is 5 years; 60 months ÷ 12 months = 5 years.

207. e. Multiply the number of minutes in an hour by the given number of hours. There are sixty minutes in an hour. Four hours is 240 minutes; 4 hours × 60 minutes = 240 minutes.

208. c. You must convert centuries to years, then years to months. First, multiply the total number of centuries by the number of years in a century. There are 100 years in a century. There are 200 years in two centuries (2 centuries × 100 years = 200 years). Now, multiply this total number of years by the number of months in a year. There are 12 months in a year. There are 2,400 months in 200 years; 200 years × 12 months = 2,400 months.

6

Conversion— Dollars and Cents

This chapter contains a series of questions that require currency conversions. You deal with money often in your everyday life. Converting between amounts of American currency is probably second nature to you. Do not get nervous, you have the knowledge and skills necessary to tackle these problems. Often, on test questions, common errors and pitfalls show up in the answer choices. Read question and answer choices carefully and always double-check your answer before making your final choice.

There are also problems that deal with foreign currency exchange rates. You will be provided with exchange rates. When you finish, review the answer explanations to see how to solve each question.

209. Two quarters are equivalent to how many dimes?

    a. 2

    b. 4

    c. 5

    d. 6

    e. 10

210. How many quarters are there in $10?

    a. 80

    b. 40

    c. 10

    d. 100

    e. 250

211. How many pennies are there in a half dollar?

    a. 2

    b. 10

    c. 25

    d. 50

    e. 100

212. A $100 bill is equal to how many $5 bills?

    a. 20

    b. 100

    c. 40

    d. 5

    e. 10

213. Three dimes and four nickels are equal to how many pennies?

    a. 7

    b. 34

    c. 70

    d. 12

    e. 50

214. Marco needs quarters for the laundry machine. How many quarters will he receive if he puts a $5 bill into the change machine?

a. 25
b. 20
c. 5
d. 4
e. 100

215. How many dimes are there in $20?

a. 20
b. 200
c. 2,000
d. 100
e. 1,000

216. How many dimes are equal to 6 quarters?

a. 16
b. 60
c. 10
d. 12
e. 15

217. Khai's meal cost $7. He gives the waitress a $20 bill and asks for the change in all $1 bills. How many $1 bills does he receive?

a. 5
b. 7
c. 2
d. 13
e. 17

218. Sixteen half-dollar coins is equal to how many dollars?

a. 8
b. 50
c. 16
d. 80
e. 100

219. How many nickels are there in three quarters?

    a. 5

    b. 25

    c. 15

    d. 6

    e. 50

220. Anna has 20 dimes. How many dollars is this?

    a. 1

    b. 2

    c. 5

    d. 10

    e. 20

221. The value of 100 quarters is equal to how many dollars?

    a. 2.50

    b. 10

    c. 25

    d. 1

    e. 50

222. Sandra used the following chart to keep track of her tips. How much money did she make in tips?

Currency	Amount of Each Currency
$1 bills	7
Quarters	10
Dimes	25
Nickels	4

    a. $12.00

    b. $15.50

    c. $15.20

    d. $12.20

    e. $10.50

223. How many dimes are there in four half-dollars?

    a. 4

    b. 10

    c. 40

    d. 200

    e. 20

224. Jack has four $50 bills. This is equal to how many $1 bills?

    a. 200

    b. 4

    c. 100

    d. 2,000

    e. 150

225. Five quarters is equal to how many nickels?

    a. 5

    b. 25

    c. 50

    d. 125

    e. 20

226. One penny, one nickel, one dime, and one quarter are equal to how much money?

    a. $0.28

    b. $0.46

    c. $0.41

    d. $0.81

    e. $1.01

227. How many quarters are there in $12?

    a. 12

    b. 120

    c. 24

    d. 84

    e. 48

228. Sonya buys a lamp that cost $25. She hands the clerk a $100 bill and asks for her change to only be in $5 bills. How many $5 bills does she receive as change?

  a. 20
  b. 15
  c. 50
  d. 25
  e. 5

229. How many pennies are there in a $20 bill?

  a. 200
  b. 2,000
  c. 20
  d. 20,000
  e. 1,000

230. A parking meter allows 15 minutes for every quarter. If Hung has $2 worth of quarters, how much time does he get on the meter?

  a. 30 minutes
  b. 45 minutes
  c. 1 hour
  d. 2 hours
  e. 4 hours

231. How many nickels are there in 5 half-dollars?

  a. 5
  b. 100
  c. 25
  d. 50
  e. 250

232. Luis won 50 nickels from the slot machine. How many dollars is this?

  a. 0.50
  b. 5
  c. 25
  d. 15
  e. 2.50

233. How many quarters are needed to equal the value of 35 dimes?

   a. 5

   b. 10

   c. 35

   d. 12

   e. 14

234. If $1 is equal to 11 pesos, how many dollars are equal to 143 pesos?

   a. 11

   b. 13

   c. 14

   d. 143

   e. 1,573

235. Dana just returned from Europe where the exchange rate was $1 = 0.80 euros. She has 56 euros left over from her trip. How many dollars is this?

   a. $44.80

   b. $70.

   c. $56.80

   d. $48.

   e. $80.

236. Twenty dollars is equal to how many pesos? ($1 = 11 pesos)

   a. 55

   b. 181

   c. 1.81

   d. 210

   e. 220

237. If a book costs 13 Canadian dollars, how many U.S. dollars is this? Assume that the current exchange rate is 1 U.S. dollar equals 1.30 Canadian dollars.

   a. $10.00

   b. $20.00

   c. $13.00

   d. $13.30

   e. $16.90

238. A British tourist needs to change 63 British pounds into U.S. dollars. If the exchange rate is 1 pound for every $1.80, how many dollars will he receive?

a. $28.50

b. $35.

c. $113.40

d. $180.

e. $65.80

239. Thirty dollars is equal to how many yen? ($1 = 107 yen)

a. 3.56

b. 28

c. 3,000

d. 3,210

e. 2,100

240. Based on the following chart of exchange rates, if Tony had 1 unit of currency from each country, which is worth the most in U.S. dollars?

Currency	The Foreign Value Equivalent of $1
Australia	1.2842
Britain	0.5416
India	45.420
Euro	0.7840

a. Australia

b. Britain

c. India

d. Euro

e. not enough information

241. How many Canadian dollars are equal to 20 U.S. dollars? (1 U.S. dollar = 1.3 Canadian dollars)

    a. 23

    b. 26

    c. 2.6

    d. 15.40

    e. 10.80

Answers

209. **c.** Each quarter is worth $0.25 and so two quarters is equal to $0.50 ($2 \times 0.25 = 0.50$). Each dime is worth $0.10; $0.50 ÷ $0.10 = 5 dimes.

210. **b.** Each quarter is $0.25, so divide $10 by this amount to get the number of quarters; $10 ÷ $0.25 = 40 quarters.

211. **d.** A half-dollar is worth 50 cents. Each penny is a cent so there are 50 pennies.

212. **a.** $100 ÷ $5 = 20. Therefore, 20 $5 bills is equal to $100.

213. **e.** 3 dimes equal 30 cents (3×10 cents); 4 nickels equal 20 cents (4×5 cents). Add these two values together to get the total; 30 + 20 = 50 cents. Therefore, this is equal to 50 pennies.

214. **b.** A quarter is worth $0.25. There are 4 quarters in every dollar ($4 \times .25 = 1$). Multiply 4 times the dollar amount to get the number of quarters; $4 \times \$5 = 20$ quarters.

215. **b.** Each dime is worth $0.10 so there are 10 dimes in every dollar ($10 \times .10 = 1$). Multiply 10 by the dollar amount to get the number of dimes; $10 \times \$20 = 200$ dimes.

216. **e.** Each quarter is worth $0.25, so 6 quarters equals $1.50 ($6 \times .25 = 1.50$). Each dime is worth $0.10. Divide $1.50 by $0.10 to get the number of dimes; 1.50 ÷ 0.10 = 15 dimes.

217. **d.** The first step is to figure out how much change he should receive; $20 – $7 = $13. Since he wants all $1 bills, he will receive 13 bills.

218. **a.** A half-dollar is equal to $0.50; $16 \times \$0.50 = \$8$.

219. **c.** Each quarter is worth 25 cents; 3×25 cents = 75 cents. Each nickel is worth 5 cents. Divide 75 cents by 5 cents to get the number of nickels; 75 ÷ 5 = 15 nickels.

220. **b.** Each dime is $0.10; 20 dimes $\times$ $0.10 = $2.00.

221. **c.** There are four quarters in every dollar; 100 quarters ÷ 4 quarters/dollar = $25.

222. **d.** First, calculate the dollar amount of each form of currency; $1
bills: $1 × 7 = $7. Quarters: $0.25 × 10 = $2.50. Dimes: $0.10
× 25 = $2.50. Nickels: $0.05 × 4 = $0.20. Next, add up all of
the dollar values to get the total; $7 + $2.50 + $2.50 + $0.20 =
$12.20.

223. **e.** Each half-dollar is worth $0.50; $0.50 × 4 = $2.00. Since each
dime is worth $0.10, divide $2.00 by this value to get the
number of dimes; $2.00 ÷ $0.10 = 20 dimes.

224. **a.** Multiply $50 by the number of bills; $50 × 4 = $200.
Therefore, this is equal to 200 $1 bills.

225. **b.** A quarter is equal to $0.25; $0.25 × 5 = $1.25; $1.25 ÷ $0.05
(the value of a nickel) = 25 nickels.

226. **c.** Add up the amounts of each coin; $0.01 (penny) + $0.05 (nickel)
+ $0.10 (dime) + $0.25 (quarter) = $0.41.

227. **e.** There are 4 quarters in every dollar; $12 × 4 quarters/dollar =
48 quarters.

228. **b.** The first step is to figure out how much change she needs back;
$100 – $25 (the cost of the lamp) = $75. Next, calculate how
many $5 bills total $75; $75 ÷ $5/bill = 15 bills.

229. **b.** There are 100 pennies in every dollar; $20 × 100
pennies/dollar = 2,000 pennies. A shortcut to this multiplication
is to multiply the whole numbers (2 × 1) and then add the
number of zeros at the end. There are three zeros (1 from the
20, two from the 100), so the total is 2,000.

230. **d.** There are four quarters in every dollar; $2 × 4 quarters/dollar =
8 quarters. Every quarter allows 15 minutes. Four quarters ×
15 min/quarter = 120 min, which is equal to two hours (60
minutes per hour).

231. **d.** A half-dollar is equal to $0.50; 5 × $0.50 = $2.50. Divide this
amount by the value of one nickel ($0.05) to find the total
number of nickels; $2.50 ÷ $0.05/nickel = 50 nickels.

232. **e.** Each nickel is worth $0.05; 50 nickels × $0.05/nickel = $2.50.

233. **e.** 35 dimes × $0.10/dime = $3.50; $3.50 ÷ $0.25/quarter = 14 quarters. An alternate solution would be to figure out that there are 4 quarters per dollar; 12 quarters would equal $3. Another 2 quarters would add the extra 50 cents. Therefore, this would equal a total of 14 quarters (12 + 2 = 14).

234. **b.** Set up the equation so that the appropriate units cancel out; 143 pesos × $1/11 pesos = $13 (143 ÷ 11 = 13 dollars).

235. **b.** Using the exchange ratio, set up the equation so that the appropriate units cancel out; 56 euros × 1/0.80 euros = $70. You must divide 56 by 0.80.

236. **e.** $20 × 11 pesos/$1 = 220 pesos. Note that the dollar units cancel out to leave the answer in units of pesos.

237. **a.** Since the ratio is $1 U.S per $1.30 Canadian, then $13 C × $1 U.S./$1.3 C = $10 U.S. Basically, divide 13 by 1.3 to get the U.S. dollar amount.

238. **c.** Multiply; 63 pounds × $1.80/1 pound = $113.40. The units of pounds cancel out so that you are left with the dollar value.

239. **d.** $30 × 107 yen/$1 = 3,210 yen. The dollar signs cancel out so that the answer is in units of yen.

240. **b.** One unit of the British currency would be worth the most in U.S. dollars. This is determined by looking for the lowest value in the chart. For example, it takes approximately 45 Indian units to equal $1. Therefore, 1 Indian unit would only be worth a small fraction of a dollar (1/45). On the other hand, 1 British pound is worth almost $2, by rounding 0.5416 to 0.5; 1 pound × $1/0.5 pounds = $2. (Divide 1 by 0.5.)

241. **b.** 20 U.S. dollars × 1.3 Canadian dollars/1 U.S. dollar = 26 Canadian dollars (20 × 1.3 = 26).

7

Conversion— Fractions, Decimals, and Percents

There are many ways to name a number, and the three most common methods are by fraction, decimal, or percent. The fractional form of a number is the most exact method of naming a number, and is often used to show a relationship, or ratio, of two numbers. The decimal form of a number is used frequently in real life situations. Our number system is a decimal number system, based on the powers of ten. Decimals are used to express monetary values, which are rounded to the nearest hundredth, or to the nearest cent. Percents are used as the most common form of ratio, where a value is compared to one hundred. Sales, taxes, and tips are all based on percentages. Percents, like decimals, are often rounded values. It is important to be able to convert numbers from one of these forms to another. The 80 questions in this chapter will test your skill at this task. Use the answer explanations to review how to perform these important conversions.

242. $\frac{3}{4}$ is equal to

 a. 0.50.

 b. 0.25.

 c. 0.75.

 d. 0.30.

 e. 1.25.

243. What is the decimal equivalent of $\frac{1}{3}$, rounded to the nearest hundredth?

 a. 0.13

 b. 0.33

 c. 0.50

 d. 0.67

 e. 0.75

244. $1\frac{1}{2}$ is equal to

 a. 0.50.

 b. 1.25.

 c. 2.50.

 d. 0.75.

 e. 1.50.

245. Maggie asked for $\frac{1}{5}$ of the cake. How much of the cake is this?

 a. 0.10

 b. 0.20

 c. 0.15

 d. 0.50

 e. 0.25

246. $\frac{4}{5}$ is equal to

 a. 0.80.

 b. 0.50.

 c. 0.90.

 d. 0.45.

 e. 1.25.

247. Ricky has completed $\frac{1}{10}$ of final project so far. How much is this in decimal form?

    a. 0.05

    b. 0.10

    c. 0.20

    d. 0.50

    e. 0.90

248. $2\frac{2}{3}$ is equal to what decimal, rounded to the nearest hundredth?

    a. 2.23

    b. 0.75

    c. 2.50

    d. 1.33

    e. 2.67

249. Diego ate $\frac{1}{5}$ his sandwich in the morning and another $\frac{1}{5}$ at lunch. How much of his sandwich did he eat in total?

    a. 0.10

    b. 0.20

    c. 0.40

    d. 0.80

    e. 1.00

250. $\frac{1}{4}$ is equal to

    a. 0.15.

    b. 0.25.

    c. 0.20.

    d. 0.75.

    e. 1.25.

251. What is $\frac{3}{8}$ equal to?

    a. 0.25

    b. 0.333

    c. 0.60

    d. 0.55

    e. 0.375

252. $\frac{11}{5}$ is equal to
   a. 2.25.
   b. 1.5.
   c. 1.15.
   d. 2.20.
   e. 0.60.

253. Jane wrote $\frac{1}{3}$ of her paper on Monday and another $\frac{1}{3}$ on Tuesday. How much of her paper does she have left to finish? Round your answer to the nearest hundredth.
   a. 0.25
   b. 0.33
   c. 0.50
   d. 0.67
   e. 0.75

Refer to the following chart to answer questions 254–255.

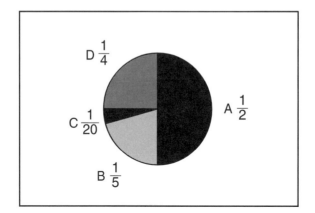

254. According to the pie chart, area B is equal to
   a. 0.50.
   b. 0.25.
   c. 0.20.
   d. 0.80.
   e. 0.05.

255. Written as a decimal, area C plus area D is equal to
   a. 0.30.
   b. 0.50.
   c. 0.35.
   d. 0.70.
   e. 0.60.

256. $3\frac{3}{10}$ is equal to
   a. 3.10.
   b. 0.30.
   c. 2.30.
   d. 3.30.
   e. 3.33.

257. Michael moved $\frac{1}{5}$ of his furniture over the weekend and another $\frac{2}{5}$ during the week. How much more does he have left to move?
   a. 0.25
   b. 0.40
   c. 0.60
   d. 0.20
   e. 0.80

258. $\frac{3}{5}$ =
   a. 0.20
   b. 0.50
   c. 0.80
   d. 0.25
   e. 0.60

259. Tia promised her younger brother that she would give him $\frac{1}{4}$ of her dollar. How much money did she give him?
   a. $0.20
   b. $0.25
   c. $0.50
   d. $0.10
   e. $0.33

260. What is $\frac{1}{10}$ in decimal form?

    a. 0.05

    b. 0.20

    c. 0.25

    d. 0.50

    e. 0.10

261. What is the decimal form of $\frac{2}{3}$, rounded to the nearest hundredth?

    a. 0.67

    b. 0.50

    c. 0.80

    d. 0.33

    e. 0.75

262. A customer at a deli orders $\frac{3}{4}$ pound of sliced turkey. This is equal to

    a. 0.25 lb.

    b. 0.30 lb.

    c. 0.75 lb.

    d. 0.80 lb.

    e. 0.33 lb.

263. What is the decimal form of $-\frac{11}{3}$, rounded to the nearest hundredth?

    a. 1.33

    b. −1.33

    c. 3.67

    d. −3.67

    e. −0.33

264. $\frac{3}{10}$ of Maria's salary goes towards taxes. She is left with how much of her original salary?

    a. 0.10

    b. 0.30

    c. 0.50

    d. 0.70

    e. 1.00

265. What is the decimal form of $\frac{5}{6}$? (Round two decimal places.)

    a. 0.65

    b. 0.88

    c. 0.83

    d. 0.13

    e. 0.56

266. Max painted $\frac{1}{2}$ of the room that he was hired to paint. His co-worker, Marie, painted $\frac{1}{2}$ of the room. What is the decimal form of the total amount painted?

    a. 0.75

    b. 0.80

    c. 0.50

    d. 1.00

    e. 0.90

267. $-\frac{1}{2} =$

    a. 0.50

    b. 1.00

    c. 0.25

    d. −0.25

    e. −0.50

268. $1\frac{3}{4} =$

    a. 1.75

    b. 0.75

    c. 1.34

    d. 1.25

    e. 3.25

Refer to the following diagram for questions 269–270.

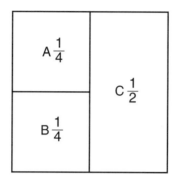

269. A landscaper is instructed to cover areas B and C with grass. This is equal to how much of the entire area?

    a. 0.25

    b. 0.50

    c. 0.80

    d. 0.60

    e. 0.75

270. The landscaper then covered half of area A with brick. In decimal form, how much of the entire area (A, B, and C) was covered with brick?

    a. 0.15

    b. 0.25

    c. 0.125

    d. 0.333

    e. 0.50

271. $3\frac{1}{4} =$

    a. 3.75

    b. 0.75

    c. 3.50

    d. 3.25

    e. 1.50

272. 0.40 =
 a. $\frac{1}{4}$
 b. $\frac{1}{5}$
 c. $\frac{2}{5}$
 d. $\frac{3}{4}$
 e. $\frac{3}{5}$

273. 0.75 =
 a. $\frac{1}{4}$
 b. $\frac{1}{5}$
 c. $\frac{2}{7}$
 d. $\frac{3}{4}$
 e. $\frac{3}{5}$

274. 2.25 =
 a. $2\frac{1}{4}$
 b. $2\frac{1}{5}$
 c. $\frac{2}{5}$
 d. $1\frac{3}{4}$
 e. $2\frac{3}{5}$

275. −0.05 =
 a. $\frac{1}{20}$
 b. $-\frac{1}{20}$
 c. $\frac{1}{2}$
 d. $-\frac{1}{2}$
 e. $\frac{5}{10}$

276. A painter needed 0.80 L of paint for a project. This is equal to
 a. $\frac{1}{4}$ L
 b. $\frac{1}{5}$ L
 c. $\frac{2}{5}$ L
 d. $\frac{3}{4}$ L
 e. $\frac{4}{5}$ L

Refer to the following table for questions 277–279.

	Portion of Assignment
Doug	0.25
Jane	0.10
A.J.	0.05
Marie	0.60

277. Doug, Jane, A.J., and Marie divided up their assigned project according to the chart. What fraction of the total project were Doug and A.J. responsible for?

   a. $\frac{1}{4}$

   b. $\frac{7}{20}$

   c. $\frac{3}{10}$

   d. $\frac{3}{4}$

   e. $\frac{2}{5}$

278. What fraction was Marie responsible for?

   a. $\frac{1}{4}$

   b. $\frac{1}{5}$

   c. $\frac{2}{5}$

   d. $\frac{3}{4}$

   e. $\frac{3}{5}$

279. If Jane dropped out of the group and A.J. took over her part of the project, what fraction was A.J. now responsible for?

   a. $\frac{1}{4}$

   b. $\frac{1}{5}$

   c. $\frac{1}{10}$

   d. $\frac{3}{20}$

   e. $\frac{1}{20}$

280. Negative 1.5 is equal to
    a. $1\frac{1}{2}$.
    b. $-1\frac{1}{5}$.
    c. $-\frac{2}{5}$.
    d. $-1\frac{1}{2}$.
    e. $\frac{3}{5}$.

281. Dat ordered 0.8 lb of smoked turkey, 0.5 lb of low-fat turkey, and 1.0 lb of honey roasted turkey from the deli. How many pounds of turkey did he order in total?
    a. $2\frac{1}{3}$
    b. $1\frac{3}{10}$
    c. $\frac{4}{5}$
    d. $1\frac{1}{2}$
    e. $2\frac{3}{10}$

282. 0.125 =
    a. $\frac{1}{25}$
    b. $\frac{1}{8}$
    c. $\frac{2}{5}$
    d. $\frac{1}{2}$
    e. $\frac{1}{5}$

283. 0.10 written as a fraction is
    a. $\frac{1}{100}$.
    b. $\frac{1}{5}$.
    c. $\frac{2}{5}$.
    d. $\frac{1}{10}$.
    e. $\frac{3}{10}$.

284. Hilary sold 0.70 of her paintings at the fair. This is equal to
    a. $\frac{1}{7}$.
    b. $\frac{7}{10}$.
    c. $\frac{2}{7}$.
    d. $\frac{7}{100}$.
    e. $\frac{3}{5}$.

285. $3.3 =$

    a. $\frac{3}{33}$

    b. $1\frac{3}{10}$

    c. $3\frac{3}{10}$

    d. $3\frac{3}{4}$

    e. $\frac{3}{3}$

Refer to the following chart for questions 286–288.

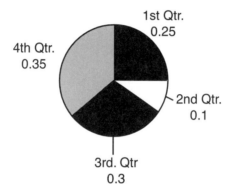

Division of Company Earnings

286. What fraction of the company's earnings was made in the third quarter?

    a. $\frac{1}{4}$

    b. $\frac{3}{10}$

    c. $\frac{1}{10}$

    d. $\frac{7}{20}$

    e. $\frac{3}{5}$

287. What fraction of the company's earnings was made in the first two quarters?

    a. $\frac{2}{5}$

    b. $\frac{3}{10}$

    c. $\frac{1}{2}$

    d. $\frac{7}{20}$

    e. $\frac{3}{5}$

288. A competing company's first-quarter earnings were double the fraction of this company's first quarter. What fraction was earned in the first quarter of the competing company?

    a. $\frac{1}{2}$

    b. $\frac{1}{4}$

    c. $\frac{2}{5}$

    d. $1\frac{1}{2}$

    e. $\frac{3}{5}$

289. $0.20 =$

    a. $\frac{1}{5}$

    b. $\frac{2}{15}$

    c. $\frac{1}{10}$

    d. $\frac{1}{4}$

    e. $\frac{1}{20}$

290. $1.90 =$

    a. $1\frac{1}{4}$

    b. $1\frac{3}{10}$

    c. $1\frac{1}{10}$

    d. $1\frac{9}{10}$

    e. $\frac{1}{90}$

291. $-0.45 =$

    a. $-\frac{1}{4}$

    b. $\frac{9}{10}$

    c. $-\frac{9}{20}$

    d. $-\frac{4}{9}$

    e. $-\frac{5}{14}$

292. Andre mowed 0.15 of the lawn in the morning and 0.50 of the lawn in the afternoon. How much more of the lawn does he have left to mow?

    a. $\frac{1}{5}$

    b. $\frac{7}{20}$

    c. $\frac{7}{15}$

    d. $\frac{2}{5}$

    e. $\frac{3}{10}$

293. A group of five friends went out to dinner and decided to split the bill evenly. If each person was expected to pay 0.20 of the total bill, what fraction should Dave pay if he was paying for both his meal and his wife's meal?

 a. $\frac{1}{4}$

 b. $\frac{3}{5}$

 c. $\frac{2}{7}$

 d. $\frac{1}{5}$

 e. $\frac{2}{5}$

294. 0.05 is equal to

 a. $\frac{1}{20}$.

 b. $\frac{1}{5}$.

 c. $\frac{1}{10}$.

 d. $\frac{5}{30}$.

 e. $\frac{1}{2}$.

295. 1.75 =

 a. $1\frac{1}{5}$

 b. $1\frac{3}{5}$

 c. $\frac{1}{75}$

 d. $1\frac{1}{4}$

 e. $1\frac{3}{4}$

296. Mia ran 0.60 km on Saturday, 0.75 km on Sunday, and 1.4 km on Monday. How many km did she run in total?

 a. $1\frac{1}{5}$ km

 b. $1\frac{3}{4}$ km

 c. $2\frac{1}{4}$ km

 d. $2\frac{3}{4}$ km

 e. $3\frac{1}{2}$ km

297. 0.80 is equal to

 a. $\frac{3}{5}$.

 b. $\frac{8}{11}$.

 c. $\frac{4}{50}$.

 d. $\frac{4}{5}$.

 e. $\frac{2}{3}$.

298. R.J. found a lamp on sale for 0.25 off its original price. What fraction of its original price will R.J. have to pay for the lamp?

  a. $\frac{1}{4}$

  b. $\frac{3}{4}$

  c. $\frac{1}{2}$

  d. $\frac{2}{3}$

  e. $\frac{4}{5}$

Refer to the following table for questions 299–301.

Amount of Paint Needed to Paint Different Pieces of Furniture

	Amount of Paint Needed
Bookcase	0.65 L
Desk	0.40 L
Chair	0.25 L

299. Jake has two chairs to paint. How much paint will he need?

  a. $\frac{1}{5}$ L

  b. $\frac{1}{2}$ L

  c. $\frac{1}{4}$ L

  d. $\frac{3}{4}$ L

  e. $\frac{1}{8}$ L

300. How much paint is needed to paint a bookcase and a desk in total?

  a. $1\frac{1}{2}$ L

  b. $\frac{49}{50}$ L

  c. $1\frac{1}{10}$ L

  d. $1\frac{1}{20}$ L

  e. 1 L

301. Chris has a dresser, chair, and desk to paint. She buys 2 L of paint. How much paint will she have leftover?

    a. $1\frac{4}{5}$ L

    b. $1\frac{2}{5}$ L

    c. $\frac{2}{5}$ L

    d. $\frac{3}{4}$ L

    e. $\frac{3}{5}$ L

302. $\frac{1}{2} =$

    a. 20%

    b. 50%

    c. 75%

    d. 25%

    e. 10%

303. What percent, rounded to the nearest percent, is $\frac{2}{3}$?

    a. 33%

    b. 50%

    c. 60%

    d. 25%

    e. 67%

304. $\frac{2}{5} =$

    a. 20%

    b. 80%

    c. 50%

    d. 40%

    e. 25%

305. By lunch, Jill had finished stocking $\frac{3}{4}$ of the shelves. What percentage is this?

    a. 75%

    b. 25%

    c. 50%

    d. 33%

    e. 80%

306. A fisherman met $\frac{2}{3}$ of his weekly goal after three days. What percentage of his goal does he have left, to the nearest percent?
   a. 25%
   b. 67%
   c. 33%
   d. 30%
   e. 60%

307. Ryan spends $\frac{7}{11}$ of his monthly salary on rent. What percentage is this? (Round to the nearest whole number.)
   a. 16%
   b. 64%
   c. 70%
   d. 11%
   e. 60%

308. $\frac{7}{20} =$
   a. 35%
   b. 7%
   c. 14%
   d. 40%
   e. 44%

Refer to the following pie chart to answer questions 309–310.

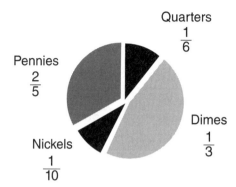

Division of Coins in Register

309. What percentage of the coins in the cash register were quarters? (Round to the nearest whole number.)

    a. 40%

    b. 10%

    c. 33%

    d. 17%

    e. 35%

310. What percentage of the coins were pennies and nickels?

    a. 40%

    b. 50%

    c. 75%

    d. 60%

    e. 15%

311. 4\5 is equal to

    a. 25%.

    b. 40%.

    c. 80%.

    d. 60%.

    e. 45%.

312. 45% is equal to

    a. $\frac{4}{5}$.

    b. $\frac{3}{5}$.

    c. $\frac{9}{20}$.

    d. $\frac{7}{20}$.

    e. $\frac{2}{5}$.

313. 25% =

    a. $\frac{1}{4}$

    b. $\frac{1}{5}$

    c. $\frac{2}{5}$

    d. $\frac{1}{2}$

    e. $\frac{3}{20}$

314. Sam bought 15% of the new stock shipment. What fraction is this?

    a. $\frac{3}{20}$

    b. $\frac{3}{5}$

    c. $\frac{7}{20}$

    d. $\frac{1}{5}$

    e. $\frac{1}{6}$

315. 60% =

    a. $\frac{1}{3}$

    b. $\frac{2}{3}$

    c. $\frac{6}{7}$

    d. $\frac{3}{5}$

    e. $\frac{2}{5}$

316. Lia finished 25% of her assignment the first week and 50% the second week. What fraction of her assignment has she done?

    a. $\frac{1}{4}$

    b. $\frac{3}{5}$

    c. $\frac{3}{4}$

    d. $\frac{1}{2}$

    e. $\frac{2}{5}$

317. A delivery man finished 20% of his deliveries in the morning and another 20% in the early afternoon. What fraction does he have left to deliver?

   a. $\frac{1}{4}$

   b. $\frac{3}{5}$

   c. $\frac{2}{5}$

   d. $\frac{1}{5}$

   e. $\frac{1}{2}$

318. 125% =

   a. $1\frac{1}{4}$

   b. $1\frac{3}{5}$

   c. $1\frac{2}{5}$

   d. $\frac{1}{5}$

   e. $1\frac{2}{5}$

319. Taylor ended up giving 20% of her profits to a charitable foundation. What fraction of her profits is this?

   a. $\frac{1}{4}$

   b. $\frac{3}{5}$

   c. $\frac{1}{3}$

   d. $\frac{1}{5}$

   e. $\frac{2}{5}$

Refer to the following table for questions 320–321.

	Percentage of Company Owned
Sandra	35%
Anna	15%
Kelly	10%
Ria	40%

320. What fraction of the company does Kelly own?

a. $\frac{2}{5}$

b. $\frac{7}{20}$

c. $\frac{1}{10}$

d. $\frac{1}{5}$

e. $\frac{1}{100}$

321. If Anna gives up her percentage to Sandra, what fraction of the company will Sandra own?

a. $\frac{1}{4}$

b. $\frac{3}{5}$

c. $\frac{2}{3}$

d. $\frac{2}{5}$

e. $\frac{1}{2}$

Answers

242. **c.** To convert a fraction to a decimal, divide the numerator, 3, by the denominator, 4; $\dfrac{0.75}{4\overline{)3.00}}$.

243. **b.** Divide the numerator by the denominator; $\dfrac{0.33\overline{3}}{3\overline{)1.000}}$.
Round the answer to the hundredths place (two decimal places) to get the answer 0.33.

244. **e.** $1\frac{1}{2}$ is a mixed number. To convert this into a decimal, first take the whole number (in this case, 1) and place it to left of the decimal point. Then take the fraction (in this case, $\frac{1}{2}$) and convert it to a decimal by dividing the numerator by the denominator; $\dfrac{0.50}{2\overline{)1.00}}$.
Putting these two steps together gives the answer, 1.50 (1.0 + 0.50 = 1.50).

245. **b.** Divide 1 by 5 in order to convert the fraction into a decimal; $\dfrac{0.20}{5\overline{)1.00}}$.

246. **a.** Divide 4 by 5 in order to convert the fraction into a decimal; $\dfrac{0.80}{5\overline{)4.00}}$.

247. **b.** Divide 1 by 10 in order to convert the fraction into a decimal; $\dfrac{0.10}{10\overline{)1.00}}$.

248. **e.** This is a mixed number so first separate the whole number from the fraction; $2\frac{2}{3} = 2.0 + \frac{2}{3}$. Convert $\frac{2}{3}$ into a decimal; $3\overline{)2.000}$ $0.66\overline{6}$.

Round the answer to the hundredths place (two decimal places), to get 0.67. Add the two numbers together to get 2.67; $(2.0 + 0.67 = 2.67)$.

249. **c.** If Diego ate $\frac{1}{5}$ of his sandwich in the morning and another $\frac{1}{5}$ at lunch, then in total, he ate $\frac{2}{5}$ of his sandwich; $\frac{1}{5} + \frac{1}{5} = \frac{2}{5}$. Divide the numerator by the denominator to get the decimal form; $5\overline{)2.00}$ 0.40.

250. **b.** Divide 1 by 4 in order to convert the fraction into a decimal; $4\overline{)1.00}$ 0.25.

251. **e.** Divide 3 by 8 in order to convert the fraction into a decimal; $8\overline{)3.000}$ 0.375.

252. **d.** $\frac{11}{5}$ is an improper fraction. One way of solving this problem is to convert the improper fraction into a mixed number which can then be converted into a decimal as in previous problems. However, a quicker method is to simply divide the numerator by the denominator, paying close attention to the decimal point; $5\overline{)11.00}$ 2.20.

253. **b.** If Jane wrote $\frac{1}{3}$ of her paper on Monday and $\frac{1}{3}$ on Tuesday, she has written $\frac{2}{3}$ of her paper; $\frac{1}{3} + \frac{1}{3} = \frac{2}{3}$. Therefore, she has $\frac{1}{3}$ left to write; $\frac{3}{3} - \frac{2}{3} = \frac{1}{3}$. Divide 1 by 3 to find the decimal form; $3\overline{)1.000}$ $0.33\overline{3}$. Round the answer to the nearest hundredth (two decimal places) to get 0.33.

254. **c.** According to the pie chart, area B is equal to $\frac{1}{5}$. Divide 1 by 5 in order to convert the fraction into a decimal; $5\overline{)1.00}$ = 0.20.

255. **a.** First convert each area separately from a fraction to a decimal. Area C = $\frac{1}{20}$ = $20\overline{)1.00}$ = 0.05.

Area D = $\frac{1}{4}$ = $4\overline{)1.00}$ = 0.25. Then add these two decimals together to get the total area; 0.05 + 0.25 = 0.30.

256. **d.** This is a mixed number so it can be broken down into the whole number plus the fraction; $3\frac{3}{10}$ = 3.0 + $\frac{3}{10}$. Divide 3 by 10 to convert the fraction to a decimal; $10\overline{)3.00}$ = 0.30. Therefore, 3.0 + 0.30 = 3.30.

257. **b.** First, figure out how much furniture he moved in total; $\frac{1}{5} + \frac{2}{5}$ = $\frac{3}{5}$. Subtract this number from $\frac{5}{5}$ to calculate how much he has left to move; $\frac{5}{5} - \frac{3}{5} = \frac{2}{5}$. The final step is to convert this fraction into a decimal by dividing 2 by 5; $5\overline{)2.00}$ = 0.40.

258. **e.** Divide 3 by 5 to convert from a fraction into a decimal; $5\overline{)3.00}$ = 0.60.

259. **b.** To find out how much money she gives her brother, divide 1 by 4; $4\overline{)1.00}$ = 0.25.

260. **e.** Divide 1 by 10 to convert the fraction into a decimal;

$$10\overline{)1.00} \quad \begin{array}{c} 0.10 \end{array}$$.

261. **a.** Divide 2 by 3 to convert the fraction into a decimal;

$3\overline{)2.000} \quad 0.66\overline{6}$. Round this answer to the nearest hundredth (two decimal places) to get 0.67.

262. **c.** To figure out the decimal equivalent, divide 3 by 4;

$4\overline{)3.00} \quad 0.75$.

263. **b.** $-1\frac{1}{3}$ is a mixed fraction and is equal to the whole number plus the fraction; $-1\frac{1}{3} = -(1 + \frac{1}{3})$. Convert $\frac{1}{3}$ into a decimal by dividing 1 by 3; $3\overline{)1.000} \quad 0.33\overline{3}$. Round this portion of the answer to the nearest hundredth, (two decimal places), to get 0.33; $-(1 + 0.33) = -1.33$.

264. **d.** First, figure out the fraction of her salary that she is left with; $\frac{10}{10} - \frac{3}{10} = \frac{7}{10}$. Then convert this to a fraction by dividing 7 by 10;

$10\overline{)7.00} \quad 0.70$.

265. **c.** Divide 5 by 6 to convert the fraction into a decimal;

$6\overline{)5.000} \quad 0.83\overline{3}$. Round two decimal places to get 0.83.

266. **d.** Max paints $\frac{1}{2}$ the room and Marie paints $\frac{1}{2}$ the room which equals 1.0 ($\frac{1}{2} + \frac{1}{2} = \frac{2}{2}$); $\frac{2}{2} = 1.0$.

267. **e.** Divide 1 by 2 in order to convert the fraction into a decimal;

$2\overline{)1.00} \quad 0.50$. Finally, add the negative sign to get −0.50.

268. **a.** The mixed number is equal to the whole number plus the fraction; $1\frac{3}{4} = 1.0 + \frac{3}{4}$. Convert the fraction to a decimal by dividing 3 by 4; $\frac{0.75}{4)\overline{3.00}}$; $1.0 + 0.75 = 1.75$.

269. **e.** First, figure out the total area of B and C; $\frac{1}{2} = \frac{2}{4}$, so $\frac{1}{4} + \frac{2}{4} = \frac{3}{4}$. Then divide 3 by 4 to get the decimal form; $\frac{0.75}{4)\overline{3.00}}$.

270. **c.** Half of area A is covered with brick. Area A $= \frac{1}{4}$, so $\frac{1}{4} \div 2 = \frac{1}{8}$. Divide 1 by 8 to convert the fraction to a decimal; $\frac{0.125}{8)\overline{1.000}}$.

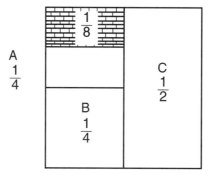

271. **d.** Separate the mixed number into the whole number plus the fraction; $3\frac{1}{4} = 3.0 + \frac{1}{4}$. Divide 1 by 4 to convert the fraction to a decimal; $\frac{0.25}{4)\overline{1.00}}$; $3.0 + 0.25 = 3.25$.

272. **c.** To convert a decimal to a fraction, first note the number of place positions to the right of the decimal point. In 0.4, the 4 is in the tenths place, which is one place to the right of the decimal point. Therefore, the fraction would be $\frac{4}{10}$. Now the fraction needs to be reduced to its lowest terms. The number 2 is the greatest common factor of 4 and 10, so divide the numerator and the denominator by 2. The final fraction is $\frac{2}{5}$.

273. **d.** In the decimal 0.75, the 75 is in the hundredths place, which is two places to the right of the decimal point. Therefore, the fraction would be $\frac{75}{100}$, which can then be reduced by dividing the top and bottom by 25, the greatest common factor of 75 and 100; $\frac{75}{100} \div \frac{25}{25} = \frac{3}{4}$

274. **a.** The number 2.25 involves a whole number, which is the 2 to the left of the decimal. This means that the answer will be a mixed number—a whole number plus a fraction. Convert the 0.25 into a fraction; $\frac{25}{100} \div \frac{25}{25} = \frac{1}{4}$. Adding the whole number, 2, to this fraction gives the answer $2\frac{1}{4}$.

275. **b.** First convert –0.05 into a fraction; $\frac{5}{100} \div \frac{5}{5} = \frac{1}{20}$. Then, do not forget to add the negative sign to get $-\frac{1}{20}$.

276. **e.** In this decimal, 0.8, the 8 is in the tenths place, so the fraction would be $\frac{8}{10}$. Reduce the fraction by dividing the top and bottom by 2 to get $\frac{4}{5}$.

277. **c.** First, add up the portion that Doug and A.J. have to do to find the total; 0.25 + 0.05 = 0.30. Then convert this into a fraction by placing 3 over 10 (since the 3 falls in the tenths place, which is one place to the right of the decimal); $\frac{3}{10}$ is the answer because this fraction cannot be reduced.

278. **e.** Look at the table to find out what portion Marie was responsible for (0.60). Convert this to a fraction and reduce; $\frac{6}{10} \div \frac{2}{2} = \frac{3}{5}$.

279. **d.** Look at the table to find out Jane's portion (0.10); Add this to A.J.'s portion (0.05). 0.10 + 0.05 = 0.15. Convert this to a fraction and then reduce; $\frac{15}{100} \div \frac{5}{5} = \frac{3}{20}$.

280. **d.** Since there is a number, 1, to the left of the decimal, this will be the whole number. The fraction is found by using the number to the right of the decimal. The 5 is in the tenths place, so the fraction is $\frac{5}{10}$ which can be reduced to $\frac{1}{2}$ (divide both the numerator and the denominator by 5). The final mixed number is the whole number (1) plus the fraction ($\frac{1}{2}$). Adding the negative, the answer is $-1\frac{1}{2}$.

281. **e.** Add up the three different values of turkey to get the total amount; $0.8 + 0.5 + 1.0 = 2.3$. Convert this to a fraction by taking the whole number, 2, which is to the left of the decimal, and add it to the fraction, $\frac{3}{10}$. The answer is $2\frac{3}{10}$.

282. **b.** In the decimal, 0.125, the 125 is in the thousandths place, since it is three places to the right of the decimal point; 125 is the greatest common factor of 125 and 1,000. The fraction is $\frac{125}{1,000} \div \frac{125}{125} = \frac{1}{8}$.

283. **d.** In the decimal, 0.10, the 1 is in the tenths place. Therefore, the fraction would be $\frac{1}{10}$, which cannot be reduced.

284. **b.** In the decimal 0.70, the 7 is in the tenths place. Place the 7 over 10 to get the fraction $\frac{7}{10}$, which cannot be further reduced.

285. **c.** Since there is a number, 3, to the left of the decimal point, this is the whole number; 0.3 is to the right of the decimal point, so this part is the fraction. The 3 is in the tenths place, so the fraction is $\frac{3}{10}$. The final mixed number answer is the whole number (3) plus the fraction ($\frac{3}{10}$); $3 + \frac{3}{10} = 3\frac{3}{10}$.

286. **b.** Refer to the pie chart to get the value for the third quarter, 0.3. Convert this to a fraction by using 10 as the denominator since the 3 in 0.3 is in the tenths place; $\frac{3}{10}$ is the answer since this fraction cannot be further reduced.

287. **d.** Refer to the pie chart and add the values for the first two quarters; $0.25 + 0.10 = 0.35$. Convert this decimal to a fraction and reduce; $\frac{35}{100} \div \frac{5}{5} = \frac{7}{20}$.

288. **a.** According to the pie chart, the first-quarter earnings were 0.25. If the competing company earned twice that amount in their first quarter, they must have earned 0.5 ($0.25 \times 2 = 0.5$). Convert this to a fraction and then reduce; $\frac{5}{10} \div \frac{5}{5} = \frac{1}{2}$.

289. **a.** In the decimal, 0.20, the 2 is in the tenths place so the fractions would be $\frac{2}{10}$. Divide both the numerator and denominator by 2 to reduce the fraction and get $\frac{1}{5}$.

290. **d.** This will be a mixed number. The whole number, 1, is the number to the left of the decimal. The fraction is the number to the right of the decimal, 9. Since the 9 is in the tenths place, the fraction would be $\frac{9}{10}$. The whole number plus the fraction is $1\frac{9}{10}$.

291. **c.** In the decimal, 0.45, the 45 is in the hundredths place (two places to the right of the decimal). Convert this to a fraction and then reduce; $\frac{45}{100} \div \frac{5}{5} = \frac{9}{20}$. Do not forget to attach the negative sign to get and answer of $-\frac{9}{20}$.

292. **b.** Add up the amounts that Andre has already mowed to get a total; $0.15 + 0.50 = 0.65$. Then subtract this from one in order to calculate how much more he has left to mow; $1.0 - 0.65 = 0.35$. Convert this to a fraction and then reduce; $\frac{35}{100} \div \frac{5}{5} = \frac{7}{20}$.

293. **e.** Each person has to pay for 0.20 of the bill, so Dave must pay twice that amount for him and his wife; $0.20 \times 2 = 0.40$. Convert this to a fraction and then reduce; $\frac{4}{10} \div \frac{2}{2} = \frac{2}{5}$.

294. **a.** In the decimal, 0.05, the 5 falls in the hundredths place (two places to the right of the decimal. To convert this to a fraction, the 5 is placed over 100 and then reduced; $\frac{5}{100} \div \frac{5}{5} = \frac{1}{20}$

295. **e.** This will be a mixed number, which is a whole number, 1, the number to the left of the decimal, and a fraction, the number to the right of the decimal. The fraction is $\frac{75}{100} \div \frac{25}{25} = \frac{3}{4}$. Adding the whole number and fraction together gives the answer $1\frac{3}{4}$.

296. **d.** Add up the individual distances to get the total amount that Mia ran; $0.60 + 0.75 + 1.4 = 2.75$ km. Convert this into a fraction by adding the whole number, 2, to the fraction: $\frac{75}{100} \div \frac{25}{25} = \frac{3}{4}$. The answer is $2\frac{3}{4}$ km.

297. **d.** In the decimal 0.80, the 8 is in the tenths place (one place to the right of the decimal). Place the 8 over 10 to get $\frac{8}{10}$ and then reduce the fraction by dividing the top and bottom by 2 to get $\frac{4}{5}$.

298. **b.** If the lamp is 0.25 off of its original price, then the sale price will be 0.75 (1.0 − 0.25 = 0.75). Convert 0.75 into a fraction and reduce; $\frac{75}{100} \div \frac{25}{25} = \frac{3}{4}$.

299. **b.** According to the table, one chair requires 0.25 L of paint. Therefore, two chairs need 0.50 L of paint; 0.25 × 2 = 0.50. Convert this to a fraction and reduce; $\frac{50}{100} \div \frac{50}{50} = \frac{1}{2}$ L.

300. **d.** Add the amounts of paint needed for the bookcase and the desk; 0.65 + 0.40 = 1.05 L. This will be a mixed number, which is a whole number, 1, plus a fraction form of 0.05. Convert 0.05 into a fraction and reduce. The 5 is in the hundredths place, so $\frac{5}{100} \div \frac{5}{5} = \frac{1}{20}$. Adding the whole number and fraction gives the answer $1\frac{1}{20}$ L.

301. **e.** The first step is to add the amounts of paint needed for the dresser, chair, and desk. 0.75 + 0.25 + 0.40 = 1.4 L. Then subtract this amount from the total amount of paint that Chris bought, 2 L; 2.0 − 1.4 = 0.6 L. This is how much paint she will have left over. Convert this number to a fraction and reduce; $\frac{6}{10} \div \frac{2}{2} = \frac{3}{5}$ L.

302. **b.** To convert from a fraction to a percentage, first convert the fraction to a decimal by dividing the numerator by the denominator; $\frac{0.5}{2)\overline{1.0}}$. Convert the decimal to a percentage by multiplying by 100; 0.5 × 100 = 50%.

303. **e.** Divide the numerator, 2, by the denominator, 3, to convert to decimal form; $\frac{0.6\overline{6}}{3)\overline{2.00}}$. Round to 0.67 and multiply by 100 to get the percentage; 0.67 × 100 = 67%.

304. **d.** Divide 2 by 5 to get the decimal form; $\dfrac{0.4}{5\overline{)2.0}}$.

Multiply this number by 100 to get the percentage; $0.40 \times 100 = 40\%$.

305. **a.** Divide 3 by 4 to get the decimal form; $\dfrac{0.75}{4\overline{)3.00}}$.

Multiply this number by 100 to get the percentage; $0.75 \times 100 = 75\%$.

306. **c.** The first step is to find out what fraction of his goal the fisherman has left; $\frac{3}{3} - \frac{2}{3} = \frac{1}{3}$. Next, convert this to decimal form by dividing the numerator, 1, by the denominator, 3; $\dfrac{0.3\overline{3}}{3\overline{)1.00}}$,

which is equal to 0.33 with rounding. Multiply this number by 100 to get the percentage; $0.33 \times 100 = 33\%$.

307. **b.** First, divide the numerator, 7, by the denominator, 11; $\dfrac{0.636}{11\overline{)7.000}}$.

Multiply this number by 100 to get the percentage; $0.636 \times 100 = 63.6\%$. Round this percentage to the nearest whole number to get 64%.

308. **a.** Divide 7 by 20 to get the decimal form; $\dfrac{0.35}{20\overline{)7.00}}$.

Multiply this number by 100 to get the percentage; $0.35 \times 100 = 35\%$.

309. d. According to the pie chart, quarters represent $\frac{1}{6}$ of the coins. Divide 1 by 6 to get the decimal form; $6\overline{)1.00} = 0.16\overline{6}$. Multiply this number by 100 and round to the nearest whole number; $0.17 \times 100 = 17\%$.

310. b. Add the fraction of pennies and nickels together to get the total; $\frac{2}{5} + \frac{1}{10} = \frac{4}{10} + \frac{1}{10} = \frac{5}{10} = \frac{1}{2}$. Divide 1 by 2 to get the decimal form; $2\overline{)1.0} = 0.5$. Multiply this number by 100 to get the percentage; $0.5 \times 100 = 50\%$.

311. c. Divide the numerator, 4, by the denominator, 5, to get the decimal form; $5\overline{)4.0} = 0.8$. Multiply this number by 100 to get the percentage; $0.8 \times 100 = 80\%$.

312. c. The first step is to convert the percentage into a decimal by dividing by 100; $45\% \div 100 = 0.45$ (move the decimal two places to the left). In the decimal, 0.45, the 45 is in the hundredths place (two places to the right of the decimal). Place 45 over 100 and reduce the fraction; $\frac{45}{100} \div \frac{5}{5} = \frac{9}{20}$.

313. a. Divide 25% by 100 to get the decimal form; $25 \div 100 = 0.25$. Place this number over 100 and reduce the fraction; $\frac{25}{100} \div \frac{25}{25} = \frac{1}{4}$.

314. a. Divide 15% by 100 to get the decimal form; $15 \div 100 = 0.15$. Place this number over 100 and reduce the fraction; $\frac{15}{100} \div \frac{5}{5} = \frac{3}{20}$.

315. d. Divide 60 by 100 to get the decimal form; $60 \div 100 = 0.60$. Place this number over 100, since it is two decimal places to the right of the decimal point, and reduce the fraction, using the greatest common factor of 20; $\frac{60}{100} \div \frac{20}{20} = \frac{3}{5}$.

316. c. Add 25% and 50% in order to figure out the total that Lia has finished thus far; 25% + 50% = 75%. Divide this number by 100 to get the decimal form; 75% ÷ 100 = 0.75. Place 75 over 100 and reduce the fraction; $\frac{75}{100} \div \frac{25}{25} = \frac{3}{4}$.

317. b. The first step is to figure out what percentage he has delivered already; 20% + 20% = 40%. Then subtract this number from 100% to figure out how much he has left to deliver; 100% − 40% = 60%. Divide this number by 100 to get the decimal form; 60 ÷ 100 = 0.60. Since the 6 is in the tenths place (one place to the right of the decimal), convert the decimal to a fraction by placing 6 over 10 and reducing the fraction; $\frac{6}{10} \div \frac{2}{2} = \frac{3}{5}$

318. a. Divide 125% by 100 in order to get the decimal form; 125% ÷ 100 = 1.25. This will be a mixed number (a whole number plus a fraction), so the number to the left of the decimal will be the whole number, 1, and the number to the right of the decimal will be made into the fraction; 0.25 converted to a fraction is $\frac{25}{100} \div \frac{25}{25} = \frac{1}{4}$. Add the whole number and the fraction together; $1 + \frac{1}{4} = 1\frac{1}{4}$.

319. d. Divide 20% by 100 to get the decimal form; 20% ÷ 100 = 0.2. Place 2 over 10 to get the fraction and reduce; $\frac{2}{10} \div \frac{2}{2} = \frac{1}{5}$.

320. c. According to the table, Kelly owns 10% of the company. Divide this number by 100 to get the decimal form; 10% ÷ 100 = 0.1. The 1 is in the tenths place, so the fraction is $\frac{1}{10}$.

321. e. Anna owns 15% of the company. Add this amount to Sandra's portion in order to get Sandra's new total; 15% + 35% = 50%. Divide this number by 100 to get the decimal form; 50% ÷ 100 = 0.5. Since the 5 is in the tenths place, the fraction will be $\frac{5}{10}$. Divide the numerator and denominator by 5 to reduce the fraction to $\frac{1}{2}$.

8

Conversion— Scientific Notation and Decimals

Scientists developed scientific notation as a method of naming numbers. In their work, scientists frequently deal with very large numbers, such as the distance from Earth to Mars, or with very small numbers, such as the diameter of an electron. This notation was invented to allow numbers such as 123,500,000,000 or 0.00000000034 to be expressed without writing either the leading or trailing zeroes. In this chapter you will encounter very large numbers, or very small numbers, and you must understand how to convert them from decimal notation to scientific notation, or from scientific notation to decimal notation.

Scientific Notation is a very specific form of expressing a number:

$$\boxed{.} \times \boxed{10}^{\square}$$

The first box represents one whole number digit that cannot be zero. The second box represents the rest of the digits, without showing any trailing zeroes. The small box to the right of the 10 is a power of 10, the exponent. The power of 10 indicates whether the number is greater than 10.0, (a positive exponent), or smaller than 1.0, (a negative exponent). The absolute value of the exponent instructs you how far the decimal point was moved in the original number to be expressed as scientific notation.

To change a number to scientific notation, first change the number to be a decimal between 1 and 10, by moving the decimal point. Show this decimal as being multiplied by a power of 10, determined by the number of placeholders that the decimal point was moved. For any original number greater than 10.0, the decimal point would have been moved left and the power (exponent) in scientific notation will be positive. For any original number less than 1.0, the decimal point would have been moved right and the power (exponent) in scientific notation will be negative.

To convert a number in scientific notation to a regular decimal number, move the decimal point the number of placeholders as dictated by the power of 10. If the power (exponent) is positive, the resulting number is greater than 10.0; therefore the decimal point will be moved to the right, and trailing zeroes may need to be included. If the power (exponent) is negative, the resulting number is less than 1.0; therefore the decimal point will be moved to the left, and leading zeroes may need to be included.

Use the following questions and answer explanations for practice on these types of problems.

322. 541,000 in scientific notation is

    a. 5.41×10^{-5}.

    b. 541×10^{4}.

    c. 5.41×10^{6}.

    d. 5.41×10^{5}.

    e. 54.1×10^{5}.

323. 580 converted to scientific notation form is

    a. 580×10^{3}.

    b. 5.80×10^{-2}.

    c. 5.8×10^{2}.

    d. 58.0×10^{2}.

    e. 58.0×10^{1}.

324. 0.000045 =

    a. 4.5×10^{-5}

    b. 4.5×10^{5}

    c. 45×10^{-6}

    d. 4.5×10^{-4}

    e. 0.45×10^{5}

325. Five million =

    a. 5.0×10^{5}

    b. 0.5×10^{4}

    c. 5.0×10^{6}

    d. 5.0×10^{-6}

    e. 5.0×10^{7}

326. 0.0908 in scientific notation form is

    a. 9.08×10^{2}.

    b. 9.08×10^{-3}.

    c. 90.8×10^{-2}.

    d. 90.8×10^{3}.

    e. 9.08×10^{-2}.

327. Juan checked his odometer and found that his car had 46,230 miles
on it. What is this number in scientific notation?
a. 462.3×10^3 miles
b. 4.623×10^5 miles
c. 4.6×10^{-5} miles
d. 4.623×10^4 miles
e. 4.623×10^{-5} miles

328. Four thousand plus nine thousand =
a. 1.3×10^4
b. 1.3×10^{-4}
c. 4.9×10^4
d. 13×10^4
e. 1.3×10^3

329. $0.00000002 =$
a. 2.0×10^9
b. 20×10^8
c. 2.0×10^{-8}
d. 2.0×10^{-7}
e. 2.0×10^8

Refer to the following chart for questions 330–331.

Company Earnings per Quarter

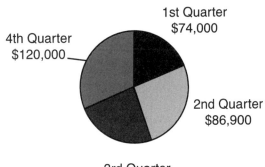

330. What is the total amount of earnings in the second quarter in scientific notation?
    a. 8.69×10^{-5}
    b. 8.69×10^{3}
    c. 86.9×10^{4}
    d. 8.69×10^{4}
    e. 8.69×10^{5}

331. What is the total amount of earnings in the first and fourth quarters together, expressed in scientific notation?
    a. 1.94×10^{5}
    b. 19.4×10^{4}
    c. 1.2×10^{6}
    d. 8.6×10^{5}
    e. 8.6×10^{-5}

332. What is 0.000009001 in scientific notation?
    a. 9.0×10^{-6}
    b. 9.0×10^{6}
    c. 9.001×10^{-6}
    d. 9.001×10^{6}
    e. $9,001 \times 10^{-9}$

333. $50 =$

    a. 5×10^{-1}

    b. 50×10^{1}

    c. 0.5×10^{3}

    d. 5×10^{1}

    e. 5×10^{2}

334. There are 1,000,000,000 nanometers in every meter. How many nanometers is this in scientific notation?

    a. 1×10^{9} nm

    b. 0.1×10^{9} nm

    c. 1×10^{10} nm

    d. 10×10^{10} nm

    e. 1×10^{8} nm

335. 2,340.25 written in scientific notation is

    a. 2.34025×10^{5}.

    b. $2,340.25 \times 10^{2}$.

    c. 2.34025×10^{3}.

    d. 2.34025×10^{2}.

    e. 2.34025×10^{4}.

336. What is -0.0063 written in scientific notation?

    a. 6.3×10^{-3}

    b. 6.3×10^{4}

    c. 63×10^{4}

    d. -6.3×10^{-3}

    e. -6.3×10^{3}

337. $8.43 \times 10^{4} =$

    a. 0.000843

    b. 0.00843

    c. 84,300

    d. 8,430,000

    e. 8,430

338. $2 \times 10^{-3} =$
   a. 0.002
   b. 0.0002
   c. 2,000
   d. 200
   e. 0.020

339. The moon is approximately 2.52×10^5 miles from the earth. What number is this equal to?
   a. 2,520
   b. 25,200,000
   c. 2,520,000
   d. 25,200
   e. 252,000

340. $1 \times 10^0 =$
   a. 10
   b. 0
   c. 1
   d. 0.1
   e. cannot be determined

341. $9.0025 \times 10^{-6} =$
   a. 9,002,500
   b. 0.000090025
   c. 0.00090025
   d. 0.0000090025
   e. 0.00000090025

342. The speed of sound is approximately 3.4×10^2 m/s at sea level. What is this number equal to?
   a. 3,400 m/s
   b. 340 m/s
   c. 34 m/s
   d. 0.0034 m/s
   e. 0.034 m/s

343. $8 \times 10^1 =$

    a. 8

    b. 0.8

    c. 80

    d. 800

    e. cannot be determined

Refer to the following chart for questions 344–345.

Note: The chart shows an alternate way of expressing scientific notation; 2.00E+03 is equivalent to 2.00×10^3.

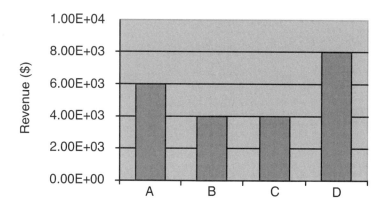

Revenue per Worker

344. What is the revenue of worker D?

    a. $600

    b. $6,000

    c. $80,000

    d. $8,000

    e. $8

345. Worker D brought in how much more revenue than worker A?

 a. $2,000

 b. $200

 c. $2

 d. $1,000

 e. $100

346. $2.11 \times 10^7 =$

 a. 2,110,000

 b. 21,100,000

 c. 211,000,000

 d. 211,000

 e. 2,110,000,000

347. $5.3 \times 10^{-1} =$

 a. 530

 b. 53

 c. 5.3

 d. 0.53

 e. 0.053

348. One μL is equal to 1×10^{-6} L. This number is equal to

 a. 0.00001 L.

 b. 0.0000001 L.

 c. 1,000,000 L.

 d. 1.000000 L.

 e. 0.000001 L.

349. A distribution manager ordered 9.35×10^4 items for the stores in the Pacific Northwest. This number is equal to

 a. 93,500.

 b. 9,350.

 c. 935,000.

 d. 935.

 e. 9,350,000.

350. $1.04825 \times 10^3 =$
   a. 1,048.25
   b. 10,482.5
   c. 104.825
   d. 104,825
   e. 104,825,000

351. $7.955 \times 10^{-5} =$
   a. 0.000007955
   b. 795,500
   c. 79,550
   d. 0.00007955
   e. 0.0007955

Answers

322. d. In this problem, the value is greater than 10.0, so the exponent on the base of 10 will be positive. The original number, 541,000, has the decimal point to the right of the last zero. Move the decimal point 5 places to the left in order to achieve the correct form for scientific notation; $541,000 = 5.41 \times 10^5$.

323. c. In order to get only one whole non-zero number (one number to the left of the decimal), move the decimal two places to the left; 580 is greater than 10.0; therefore the exponent is positive 2.

324. a. First, find the decimal place that will leave one digit to the left of the decimal (4.5). The decimal point needs to be moved 5 places to the right in order to get there. Since it is being moved to the right, and the number is smaller than 1.0, the power will be negative; 4.5×10^{-5}.

325. c. Five million is equal to 5,000,000. In order to convert this into scientific notation, the decimal needs to be moved 6 places to the left to get 5×10^6. The power is positive because the original number is greater than 10.0.

326. e. Move the decimal two places to the right to get one whole number. Add a negative sign to the power of 10 since the decimal is moved to the right, and the number is smaller than 1.0, 9.08×10^{-2}.

327. d. Move the decimal 4 places to the left in order to get only one digit to the left of the decimal; 4.623×10^4. The power is positive because the original number is greater than 10.0.

328. a. The first step is to add up the two numbers; $4,000 + 9,000 = 13,000$. Now convert to scientific notation by moving the decimal 4 places to the left; 1.3×10^4; 13,000 is greater than 10.0, so the exponent is positive.

329. c. Carefully count how many places the decimal must be moved to the right in order to end up with the number 2.0 (8 places to the right). The power sign is negative since the decimal is moved to the right; 2.0×10^{-8}. The original number is less than 1.0, so the exponent is negative.

330. **d.** According to the chart, $86,900 was made in the second quarter. Move the decimal 4 places to the left; 8.69×10^4; 86,900 is greater than 10.0, so the exponent is positive.

331. **a.** Add together the earnings from the first and fourth quarters; $74,000 + $120,000 = $194,000. Then move the decimal point five places to the left; 1.94×10^5. The sum of the earnings is greater than 10.0, so the exponent is positive.

332. **c.** Move the decimal point six places to the right in order to convert this number to scientific notation. Add the negative sign to the power since the decimal is moved to the right, and the number is smaller than 1.0. So, the answer is 9.001×10^{-6}.

333. **d.** Move the decimal one place to the left in order to have only one digit to the left of the decimal; 5×10^1. The exponent is positive because 50 is greater than 10.0.

334. **a.** In order to get to the number 1.0, the decimal needs to be moved 9 places to the left. Therefore, the amount of nanometers in a meter is 1×10^9. The exponent is positive because the original number is greater than 10.0.

335. **c.** In this problem, do not be confused by the fact that the decimal does not start off at the end of the number. The decimal needs to be moved three places to the left to get 2.34025×10^3. The original value is greater than 10.0, so the exponent is positive.

336. **d.** Move the decimal 3 places to the right in order to get the correct scientific notation form. Add the negative sign to the power since the decimal is moved to the right, and the number is smaller than 1.0. Finally, add the negative sign in front of the original number because this has not changed, it is still a negative value. So, the answer is -6.3×10^{-3}.

337. **c.** The power (exponent) is positive 4, so the number is a value greater than 10, and the decimal point must be moved four places to the right. Two trailing zeroes must be added, to get the number 84,300.

338. **a.** The power sign is negative 3, so the number will be less than 1.0, and the decimal needs to be moved three places to the left. Do not forget that 2 is equal to 2.0 (although the decimal is not shown, it follows the last digit). The answer is 0.002.

339. **e.** The power is positive 5, so the number will be greater than 10.0, and the decimal is moved five places to the right to become 252,000.

340. **c.** Any number to the zero power is equal to 1. Therefore, $10^0 = 1$. Placing this back into the problem gives 1×1 which equals 1.

341. **d.** The power is negative 6, so the number will be less than 1.0, and the decimal is moved six places to the left to get 0.0000090025.

342. **b.** The power is positive 2, so the number will be greater than 10.0, and the decimal is moved two places to the right to give the answer, 340 m/s.

343. **c.** The power is positive 1, so the number will be greater than 10.0, and the decimal is moved one place to the right to give the answer, 80.

344. **d.** According to the graph, the revenue of worker D is 8.00E+03 which is 8.0×10^3. The power is positive 3, so the dollar amount is greater than 10.0, and the decimal is moved three places to the right to give the answer of $8,000.

345. **a.** Worker D has a revenue of 8.00E+03 (8.0×10^3). Worker A has a revenue of 6.00E+03 (6.0×10^3). Both of these numbers have power signs of positive 3, so the monetary values are greater than 10.0, and the decimals are moved three places to the right to give $8,000 (D) and $6,000 (A). Subtract A from D; $8,000 − $6,000 = $2,000.

346. **b.** The power is positive 7, so the number will be greater than 10.0, and the decimal is moved seven places to the right in order to get 21,100,000.

347. **d.** The power is negative 1, so the number will be less than 1.0, and the decimal is moved one place to the left in order to get 0.53.

348. **e.** The power is negative 6, so the number will be less than 1.0, and the decimal is moved six places to the left. Therefore, one μL is equal to 0.000001 L.

349. **a.** The power is positive 4, so the number will be greater than 10.0, and the decimal is moved four places to the right to give the answer 93,500.

350. **a.** The power is positive 3, so the number will be greater than 10.0, and the decimal is moved three places to the right, even if this still leaves some digits to the right of the decimal point. The answer is 1,048.25.

351. **d.** The power is negative 5, so the number will be less than 1.0, and the decimal is moved five places to the left in order to get the answer, 0.00007955.

9

Conversion—
Metric and Standard Units:
Length, Area, and Volume

In chapters 1, 2, and 3 of this book, the topics of measuring lengths, areas, and volumes were covered. This section reviews these measurements in the context of the conversion of units, both standard and metric. Unit conversions for temperature and weight are covered in the next chapter of this book.

Conversions within the metric system are relatively easy because there are fundamental units, such as meter and liter. All other units are based on these fundamental units and the powers of 10. Conversions within the standard system require more memorization of how the units are related to one another. The formulas in this chapter outline the two systems of measurement and the conversion factors needed to convert both within one system, and from one system to the other. The rounding of answers is almost always required when converting between these systems. There are 90 questions in this chapter that will challenge you to master these conversions. The formulas and the answer explanations are provided to reinforce your skills.

Formulas

Conversions between standard units

1 foot (ft) = 12 inches (in)
1 yard (yd) = 3 feet = 36 inches
1 mile (mi) = 1,760 yards = 5,280 feet

1 square foot (ft^2) = 144 square inches (in^2)
1 square yard (yd^2) = 9 square feet
1 square mile (mi^2) = 640 acres = 3,097,600 square yards
1 acre (A) = 4,840 square yards = 43,560 square feet

1 cubic foot (ft^3) = 1,728 cubic inches (in^3) = 7.48 gallons
1 cubic yard (yd^3) = 27 cubic feet

1 tablespoon = 0.5 fluid ounce
1 tablespoon = 3 teaspoons
1 cup (c) = 8 fluid ounces (fl oz)
1 pint (pt) = 2 cups = 16 fluid ounces
1 quart (qt) = 2 pints = 32 fluid ounces
1 gallon (gal) = 4 quarts = 128 fluid ounces
1 gallon ≈ 231 cubic inches ≈ 0.1337 cubic foot

Conversions between metric units

1 centimeter (cm) = 10 millimeters (mm)
1 meter (m) = 100 centimeters = 1,000 millimeters
1 kilometer (km) = 1,000 meters

1 square centimeter (cm^2) = 100 square millimeters (mm^2)
1 square meter (m^2) = 10,000 square centimeters = 1,000,000 square millimeters
1 square kilometer (km^2) = 1,000,000 square meters

1 cubic centimeter (cm^3) = 1,000 cubic millimeters (mm^3)
1 cubic meter (m^3) = 1,000,000 cubic centimeters

1 liter (L) = 1,000 milliliters (mL) = 100 centiliters (cL)
1 kiloliter (kL) = 1,000 liters = 1,000,000 milliliters

Conversions between standard and metric units

1 inch ≈ 25.4 millimeters ≈ 2.54 centimeters
1 foot ≈ 0.3048 meter ≈ 30.480 centimeters
1 yard ≈ 0.9144 meter
1 mile ≈ 1,609.34 meters ≈ 1.6093 kilometers
1 kilometer ≈ 0.6214 mile
1 meter ≈ 3.281 feet ≈ 39.37 inches
1 centimeter ≈ 0.3937 inch

1 square inch ≈ 645.16 square millimeters ≈ 6.4516 square centimeters
1 square foot ≈ 0.0929 square meter
1 square yard ≈ 0.8361 square meter
1 square mile ≈ 2,590,000 square meters ≈ 2.59 square kilometers
1 acre ≈ 4,046.8564 square meters ≈ 0.004047 square kilometers

1 cubic inch ≈ 16,387.064 cubic millimeters ≈ 16.3871 cubic centimeters
1 cubic foot ≈ 0.0283 cubic meter
1 cubic yard ≈ 0.7646 cubic meter

1 teaspoon ≈ 5 milliliters
1 tablespoon ≈ 15 milliliters
1 fluid ounce ≈ 29.57 milliliters ≈ 2.957 centiliters
1 fluid ounce ≈ 0.00002957 cubic meters
1 gallon ≈ 3.785 liters
1 liter ≈ 1.057 quarts ≈ 0.264 gallon
1 quart ≈ 0.946 liter

352. 6 ft = _____ yd

    a. 0.5

    b. 2

    c. 3

    d. 9

    e. 18

353. A recipe calls for 3 ounces of olive oil. Convert this measurement into cups.

    a. 0.3 c

    b. 0.5 c

    c. 0.375 c

    d. 24 c

    e. $2\frac{2}{3}$ c

354. 48 in = _____ yd

    a. $0.\overline{33}$

    b. 1

    c. $1.\overline{33}$

    d. 2

    e. 84

355. A 2-liter bottle of soda contains approximately how many fluid ounces?

    a. 0.06 fl oz

    b. 968.96 fl oz

    c. 128.53 fl oz

    d. 256 fl oz

    e. 67.6 fl oz

356. The perimeter of a room is measured and found to be 652 inches. Trim for the room is sold by the foot. How many feet of trim must be purchased so that the room can be trimmed?

    a. 50 ft

    b. 54 ft

    c. 55 ft

    d. 60 ft

    e. 18 ft

357. Thomas is 6 feet 1 inch in height. His son is 3 feet 3 inches tall. What is the difference in their heights, in inches?
   a. 30 in
   b. 32 in
   c. 34 in
   d. 36 in
   e. 38 in

358. Martha walks to school, a distance of 0.85 miles. What is the distance she walks to school in feet?
   a. 4,488 feet
   b. 6,212 feet
   c. 1,496 feet
   d. 5,280 feet
   e. 1,760 feet

359. A road race is 33,000 feet long. How many miles long is the race?
   a. 18.75 mi
   b. 6.25 mi
   c. 11,000 mi
   d. 38,280 mi
   e. 5 mi

360. A child's sandbox is being constructed in Tony's backyard. The sandbox is 6 feet wide, and 5 feet long. Tony wants the sand to be at least 1.5 feet deep. The volume of sand in the box is 6 feet × 5 feet × 1.5 feet = 45 cubic feet. Convert the volume into cubic yards.
   a. 72 cubic yards
   b. 15 cubic yards
   c. 0.6 cubic yards
   d. $1.\overline{66}$ cubic yards
   e. 5 cubic yards

361. Which of the following represents a method by which one could convert inches into miles?

a. multiply by 12, then multiply by 5,280
b. divide by 12, then divide by 5,280
c. add 12, then multiply by 5,280
d. multiply by 12, then divide by 5,280
e. divide by 12, then multiply by 5,280

362. 4.5 mi = _____ ft

a. 13.5
b. 5,275.5
c. 5,284.5
d. 7,920
e. 23,760

363. 2 pints 6 ounces + 1 cup 7 ounces =

a. 1 quart
b. 3 pints 1 cup 13 ounces
c. 2 pints 5 ounces
d. 3 pints 5 ounces
e. 3 pints 1 cup

364. 35 mm = _____ cm

a. 0.35
b. 3.5
c. 35
d. 350
e. 3,500

365. Susan wishes to create bows from 12 yards of ribbon. Each bow requires 6 inches of ribbon to make. How many inches of ribbon does Susan have?

a. 18 in
b. 432 in
c. 48 in
d. 144 in
e. 24 in

366. The living room in Donna's home is 182 square feet. How many square yards of carpet should she purchase to carpet the room?

a. 9 yd^2

b. 1,638 yd^2

c. 61 yd^2

d. 21 yd^2

e. 546 yd^2

367. Sofie needed to take $\frac{3}{4}$ teaspoon of cough syrup 3 times a day. Convert $\frac{3}{4}$ teaspoon into milliliters.

a. 0.375 milliliter

b. 3.75 milliliters

c. 2.25 milliliters

d. 22.5 milliliters

e. 0.15 milliliter

368. 3.9 kiloliters = _____ milliliters

a. 0.00000039

b. 0.0000039

c. 0.0039

d. 3,900,000

e. 39,000,000

369. The 1,500-meter race is a popular distance running event at track meets. How far is this distance in miles?

a. 0.6214 mi

b. 0.5 mi

c. 0.9321 mi

d. 2.4139 mi

e. 0.4143 mi

370. 58.24 mm^3 = _____ cm^3

a. 0.05824

b. 5.824

c. 58,240

d. 582.4

e. 5,824

371. Abigail is 5.5 feet tall. Which of the following is closest to this height?

    a. 17 m

    b. 170 cm

    c. 1.7 km

    d. 170 mm

    e. 0.17 km

372. 62.4 m ≈ _____ ft

    a. 1.58

    b. 2,456.7

    c. 65.7

    d. 19.01

    e. 205

373. Which of the following statements is false?

    a. 32 cm = 320 mm

    b. 3.2 m = 3,200 mm

    c. 84 mm = 840 cm

    d. 84 mm = 0.084 m

    e. 8.4 km = 8,400 m

374. 350 m^2 = _____ km^2

    a. 350,000,000

    b. 350,000

    c. 3.5

    d. 0.000350

    e. 0.000000350

375. A Ford Mustang has a 5.0-liter engine. Write this value in gallons.

    a. 0.264 gal

    b. 1.32 gal

    c. 5.264 gal

    d. 3.79 gal

    e. 18.94 gal

376. $0.16 \, A =$ _____ yd^2
    a. 774.4
    b. 6,969.6
    c. 30,250
    d. 4,840.16
    e. 4,839.84

377. $4.236 \, km + 23 \, m + 654 \, cm =$ _____ m
    a. 681.236
    b. 33.776
    c. 4,913
    d. 453.14
    e. 4,265.54

378. Murray owns 5 square kilometers of farmland in Pennsylvania. How many acres does Murray own?
    a. 0.020235 A
    b. 1,235.48 A
    c. 20.234 A
    d. 1.235 A
    e. 2.0234 A

379. $1 \, m^3 \approx$ _____ ft^3
    a. 0.0283
    b. 3.534
    c. 35.34
    d. 2.83
    e. 28.3

380. A pool holds 14,360 gallons of water. Approximately how many cubic feet of water is this? Round to the nearest cubic foot.
    a. 107,413 ft^3
    b. 8 ft^3
    c. 1,920 ft^3
    d. 14,367 ft^3
    e. 14,353 ft^3

381. A stretch limousine is 4.5 meters long. How many yards long is the limousine?
   a. 7.5 yd
   b. 44.307 yd
   c. 13.5 yd
   d. 14.769 yd
   e. 4.92 yd

382. 12.59 m^2 ≈ _____ yd^2.
   a. 15.06
   b. 10.53
   c. 13.43
   d. 11.75
   e. 1.17

383. Sandi purchased 3 cubic yards of mulch for her garden. About how many cubic meters of mulch did she buy?
   a. 2.2304 m^3
   b. 3.7696 m^3
   c. 3.9236 m^3
   d. 2.2938 m^3
   e. 3.5621 m^3

384. 15 cL ≈ _____ fl oz
   a. 44.36
   b. 17.96
   c. 5.1
   d. 4.3
   e. 12.04

385. The average adult should drink eight 8-oz glasses of water per day. Convert the daily water requirement to gallons.
   a. 0.64 gal
   b. 6.4 gal
   c. 0.5 gal
   d. 5 gal
   e. 3.2 gal

386. The diameter of a plain m&m® is approximately 0.5 in. About how many m&m® candies would you have to place end to end to form a line a mile long?

    a. 63,360

    b. 126,720

    c. 10,560

    d. 880

    e. 31,680

387. The State of Connecticut covers 5,544 square miles of the earth's surface. How many acres is this?

    a. 4,904 acres

    b. 6,184 acres

    c. 3,103,144 acres

    d. 8.6625 acres

    e. 3,548,160 acres

388. 19 qt = _____ gal

    a. 76

    b. 4.75

    c. 9.5

    d. 2.375

    e. 1.1875

389. A tree is measured and found to be 16.9 meters tall. How many centimeters tall is the tree?

    a. 0.169 cm

    b. 1.69 cm

    c. 169 cm

    d. 1,690 cm

    e. 16,900 cm

390. 1 quart + 1 pint = _____ liter(s)

    a. 1.419 liters

    b. 1.586 liters

    c. 2.446 liters

    d. 1.5 liter

    e. 0.946 liter

391. A science experiment calls for the use of a porous sponge approximately 35 cubic centimeters in volume. About how big is the sponge in cubic inches?

a. 4.52 in³

b. 5,735.49 in³

c. 4,682.03 in³

d. 573.55 in³

e. 2.14 in³

392. The Johnson's kitchen is 221 square feet in size. How many square meters of tile must they purchase to tile the kitchen floor?

a. 22.84 m²

b. 20.53 m²

c. 2,378.90 m²

d. 221.09 m²

e. 23.91 m²

393. 345.972 mL = _____ L

a. 345,972

b. 3,459.72

c. 34.5972

d. 3.45972

e. 0.345972

394. How many cubic inches are in a cubic yard?

a. 64 in³

b. 46,656 in³

c. 1,755 in³

d. 108 in³

e. 62,208 in³

395. _____ fl oz = $2\frac{3}{4}$ c

a. 24

b. 16

c. $10\frac{3}{4}$

d. 18.72

e. 22

396. Brenda measures her kitchen window so she can purchase blinds. She finds that the window is 40 inches long. Convert this dimension to centimeters.
   a. 101.6 centimeters
   b. 15.75 centimeters
   c. 157.5 centimeters
   d. 1,016 centimeters
   e. 42.54 centimeters

397. A photo has an area of 300 square centimeters. Convert this area into square millimeters.
   a. 3,000 square millimeters
   b. 30,000 square millimeters
   c. 300,000 square millimeters
   d. 3 square millimeters
   e. 0.3 square millimeters

398. 34 yd. ≈ _____ m
   a. 31.09
   b. 3.109
   c. 37.18
   d. 3.718
   e. 34.9144

399. Many sodas are sold in 2-liter bottles. Convert 2 liters into gallons.
   a. 5.28 gallons
   b. 7.57 gallons
   c. 0.528 gallon
   d. 1.8925 gallons
   e. 0.757 gallon

400. _____ in^2 = 13 ft^2
   a. 11.08
   b. 0.09
   c. 18.72
   d. 1,872
   e. 1,108

401. A signpost is 4.3 meters tall. How tall is the signpost in centimeters?

  a. 0.043 centimeters

  b. 0.43 centimeters

  c. 43 centimeters

  d. 430 centimeters

  e. 4,300 centimeters

402. Use the diagram to answer the question:

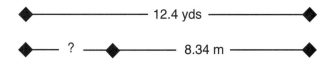

Find the missing length in meters.

  a. 4.06 meters

  b. 1.356 meters

  c. 1.134 meters

  d. 9.12 meters

  e. 3 meters

403. A book cover has an area of 72 square inches. What is its area in square centimeters?

  a. 46.45 square centimeters

  b. 464.5 square centimeters

  c. 11.16 square centimeters

  d. 111.6 square centimeters

  e. 4,645 square centimeters

404. $11.8 \text{ in}^3 = $ _____ mm^3

  a. 19,336.7

  b. 16,398.86

  c. 1,388.73

  d. 193,367.36

  e. 163,988.6

405. Fred wants to replace the felt on his pool table. The table has an area of 29,728 square centimeters. About how many square meters of felt does he need?

    a. 30 square meters

    b. 20 square meters

    c. 10 square meters

    d. 3 square meters

    e. 2 square meters

406. A storage bin can hold up to 3.25 cubic meters of material. How many cubic centimeters of material can the bin accommodate?

    a. 325,000 cubic centimeters

    b. 3,250,000 cubic centimeters

    c. 32,500,000 cubic centimeters

    d. 32,500 cubic centimeters

    e. 0.00000325 cubic centimeters

407. William drank 9 cups of juice on Saturday. How many fluid ounces of juice did he drink?

    a. 17 fluid ounces

    b. 11.25 fluid ounces

    c. 7.2 fluid ounces

    d. 1.125 fluid ounces

    e. 72 fluid ounces

408. A birdbath can hold about 226 cubic inches of water. Convert this amount into cubic feet.

    a. 0.13 ft^3

    b. 1.3 ft^3

    c. 13 ft^3

    d. 7.65 ft^3

    e. 0.765 ft^3

409. A pitcher of punch contains 1.5 liters. Convert this amount to kiloliters.

    a. 15 kiloliters

    b. 1,500 kiloliters

    c. 0.0015 kiloliters

    d. 0.15 kiloliters

    e. 1.5 kiloliters

410. 13 pints = _____ quarts
   a. 26
   b. 3.25
   c. 6.5
   d. 52
   e. 17

411. A pen is 13 centimeters long. How long is the pen in millimeters?
   a. 13 mm
   b. 130 mm
   c. 1,300 mm
   d. 0.13 mm
   e. 13,000 mm

412. 6,540 milliliters = _____
   a. 0.00654 liters
   b. 6.54 kiloliters
   c. 22.14 fluid ounces
   d. 6.54 liters
   e. 2.214 fluid ounces

413. It is 4.21 miles to the nearest mall. Convert this distance to kilometers.
   a. 6.775 km
   b. 67.75 km
   c. 0.6775 km
   d. 2.616 km
   e. 0.2616 km

414. A bottle of seltzer holds 10.5 fluid ounces. This is equivalent to how many milliliters?
   a. 0.36 milliliters
   b. 281.65 milliliters
   c. 2.82 milliliters
   d. 31.049 milliliters
   e. 310.49 milliliters

415. A dripping faucet has filled a pot that can hold 450 cubic inches of liquid. How many gallons of water have dripped from the faucet?
    a. 0.1948 gallons
    b. 2.32 gallons
    c. 1.948 gallons
    d. 0.51 gallons
    e. 5.1 gallons

416. 4,876.2 m = _____ km
    a. 0.48762
    b. 4.8762
    c. 48.762
    d. 4,872
    e. 4,876,200

417. 4.45 km^2 = _____ m^2
    a. 4,450,000
    b. 4,450
    c. 0.00445
    d. 0.00000445
    e. 445

418. How many liters (L) of coffee are in the mug?

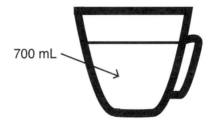

700 mL

    a. 0.7 liters
    b. 7 liters
    c. 0.07 liters
    d. 70 liters
    e. 0.007 liters

419. Michael measured a worm he found in his backyard. He discovered it was 42 millimeters long. About how long is the worm in inches?

    a. 16.54 inches

    b. 0.605 inches

    c. 6.05 inches

    d. 1.65 inches

    e. 16.6 inches

420. A container with a capacity of 3 cubic meters can hold how many fluid ounces?

    a. 88.71 fluid ounces

    b. 887 fluid ounces

    c. 0.0000887 fluid ounces

    d. 10,145.4 fluid ounces

    e. 101,454 fluid ounces

421. Refer to the diagram of a tabletop below to answer the question:

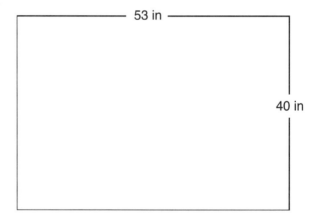

Find the area of the table in square feet.

    a. 1.767 square feet

    b. 17.67 square feet

    c. 176.7 square feet

    d. 1.472 square feet

    e. 14.72 square feet

422. A test tube is marked in cubic centimeters. If the test tube is filled to the 24 cubic centimeter mark with water, how many cubic millimeters of water are in the tube?
a. 2,400 cubic millimeters
b. 240 cubic millimeters
c. 24,000 cubic millimeters
d. 0.024 cubic millimeters
e. 2.4 cubic millimeters

423. 7.98 L = _____ mL
a. 0.798
b. 0.00798
c. 79,800
d. 7,980
e. 798,000

424. 765 ft³ ≈ _____ yd³
a. 28.33
b. 255
c. 85
d. 2.833
e. 2.55

425. A local charity has asked a water distributor to donate water for participants to drink during a fundraising walk. The company donated 231 gallons of water. How many quarts is this?
a. 57.75 quarts
b. 462 quarts
c. 115.5 quarts
d. 1,848 quarts
e. 924 quarts

426. Fill in the missing value in the table below:

Kilometers	Meters	Centimeters	Millimeters
2.8	?	280,000	2,800,000

a. 280
b. 2,800
c. 28,000
d. 28
e. 0.0028

427. A cargo ship is loaded with 12 kiloliters of water for donation to a region destroyed by an earthquake. The water costs $0.05 per liter to ship. What is the cost of the shipment?
a. $6
b. $60
c. $600
d. $6,000
e. $60,000

428. Spring water comes in 2.5-gallon containers for dispensing directly from the refrigerator. How many liters can a 2.5-gallon container hold?
a. 6.25 liters
b. 1.514 liters
c. 9.4625 liters
d. 0.946 liters
e. 0.6605 liters

429. The County Fairgrounds in Newbrite cover 3 square miles of farmland. How many square kilometers do the fairgrounds cover?
a. 7.77 square kilometers
b. 1.1583 square kilometers
c. 0.8633 square kilometers
d. 5.59 square kilometers
e. 6.24 square kilometers

430. Francis purchased enough powdered formula to make 34 quarts of liquid for her infant son. How many liters of formula will the powder make?

a. 36.784 liters

b. 35.938 liters

c. 34.946 liters

d. 32.164 liters

e. 35.94 liters

431. Use the diagram to answer the question:

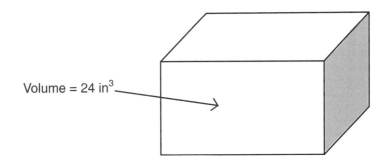

Volume = 24 in^3

Find the volume of the box in cubic millimeters (mm^3).

a. 393.294 mm^3

b. 0.001464 mm^3

c. 682.794 mm^3

d. 393,289.536 mm^3

e. 39,328.95 mm^3

432. Put in order from largest to smallest:
540 m; 7,500 cm; 0.5 km; 4,000 mm

a. 7,500 cm; 4,000 mm; 540 m; 0.5 km

b. 540 m; 0.5 km; 7,500 cm; 4,000 mm

c. 4,000 mm; 7,500 cm; 0.5 km; 540 m

d. 540 m; 4,000 mm; 0.5 km; 7,500 cm

e. 540 m; 0.5 km; 4,000 mm; 7,500 cm

433. Mary has a scale model of a train set, complete with detailed settings and buildings. If the area of the train platform in her model is 4,500 square millimeters, what is the area in square centimeters?

   a. 450 square centimeters
   b. 45,000 square centimeters
   c. 4.5 square centimeters
   d. 450,000 square centimeters
   e. 45 square centimeters

434. Which of the following is the smallest?
20 L, 0.2 kL, or 200,000 mL

   a. 20 L
   b. 0.2 kL
   c. 200,000 mL
   d. They are all equal.
   e. The answer cannot be determined from the information given.

435. 75 c = _____ pts

   a. 9.375
   b. 18.75
   c. 37.5
   d. 150
   e. 300

436. Lauri and her friend Iona live 2,500 meters apart. How far apart are they in miles? Round the answer to the nearest tenth of a mile.

   a. 1.6 miles
   b. 0.6 miles
   c. 890.7 miles
   d. 2.1 miles
   e. 1.7 miles

437. Mona is making coffee for her guests. She has 2.2 liters of coffee and wants to know how many cups that is so she can decide whether she needs to make more. Convert 2.2 liters into cups.

    a. 4.6508 cups

    b. 9.3016 cups

    c. 2.3254 cups

    d. 8.3254 cups

    e. 13.028 cups

438. _____ m^3 = 9,760, 000 cm^3

    a. 9.76

    b. 97.6

    c. 976

    d. 9,760

    e. 9,760,000,000

439. Jacob boasted to his friends, "I guzzled 1,500 milliliters of water at one time!" How many liters of water did Jacob guzzle?

    a. 15 liters

    b. 0.15 liters

    c. 1,500,000 liters

    d. 1.5 liters

    e. 15,000 liters

440. _____ m ≈ 4,321 ft

    a. 1,440

    b. 1,231

    c. 14,181

    d. 14,177

    e. 1,317

441. A warehouse can accommodate up to 4,000,000 cubic meters of inventory. How many cubic centimeters can the warehouse hold?

    a. 4,000,000,000 cubic centimeters

    b. 4,000 cubic centimeters

    c. 4,000,000,000,000 cubic centimeters

    d. 4 cubic centimeters

    e. 40,000,000,000 cubic centimeters

Answers

352. **b.** Every 3 feet equals 1 yard, so divide: 6 ft ÷ 3 ft/yd = 2 yds.

353. **c.** Every 8 ounces equals 1 cup, so divide: 3 oz ÷ 8 oz/c = 0.375 c.

354. **c.** Every 36 inches equals 1 yard, so divide: 48 in ÷ 36 in/yd = $1\frac{1}{3}$ yds, or $1.\overline{33}$ written as a decimal.

355. **e.** Each liter equals 0.264 gallons, so for 2 liters: 2 L (0.264 gal/L = 0.528 gal. A gallon equals 128 fluid ounces, so multiply: 0.528 gal × 128 fl oz/gal = 67.584 fl oz ≈ 67.6 fl oz.

356. **c.** 12 inches equals 1 foot, so divide: 652 in ÷ 12 in\ft = 54.333 feet. Since trim is sold by the foot, round up; 55 feet must be purchased so that there is enough trim.

357. **c.** One foot equals 12 inches so Thomas's height is 6 ft × 12 in\ft = 72 inches + 1 inch = 73 inches. His son's height is 3 ft × 12 in/ft = 36 inches + 3 inches = 39 inches. The difference between their heights is 73 in − 39 in = 34 inches.

358. **a.** A mile is 5,280 feet, so to find 0.85 mile, multiply: 0.85 mi × 5,280 ft/mi = 4,488 feet.

359. **b.** It takes 5,280 feet to make a mile. To find how many miles are in 33,000 feet, divide: 33,000 ft ÷ 5,280 ft/mi = 6.25 mi.

360. **d.** One cubic yard requires 27 cubic feet. To find the number of cubic yards in 45 cubic feet, divide: 45 ft^3 ÷ 27 ft^3/yd^3 = $1.\overline{66}$ yd^3.

361. **b.** Since the conversion is from smaller units to larger units, division is required. Every 12 inches equals 1 foot, so to figure out the number of feet in the given number of inches, divide by 12. Then, to figure out how many miles are in the calculated number of feet, divide by the number of feet in a mile, 5,280.

362. **e.** Each mile equals 5,280 feet. Since there are 4.5 miles, multiply: 4.5 mi × 5,280 ft/mi = 23,760 ft.

363. **d.** The two amounts can be added as they are, but then the sum needs to be simplified; 2 pints 6 ounces + 1 cup 7 ounces = 2 pints 1 cup 13 ounces. Note that the sum is written with the largest units first and smallest units last. To simplify, start with the smallest unit, ounces, and work toward larger units. Every 8 ounces makes 1 cup, so 13 ounces = 1 cup 5 ounces. Replace the 13 ounces with the 1 cup 5 ounces, adding the cup portions together: 2 pints 2 cups 5 ounces. Now note that 2 cups = 1 pint, so the sum can be simplified again combining this pint with the 2 pints in the sum. The simplified total is 3 pints 5 ounces.

364. **b.** There are 10 millimeters in every centimeter, so divide: 35 mm ÷ 10 mm/cm = 3.5 cm.

365. **b.** There are 36 inches in a yard. To find the number of inches in 12 yards, multiply: 12 yd × 36 in/yd = 432 in.

366. **d.** It takes 9 square feet to make a square yard. To find out how many square yards are in 182 square feet, divide: 182 ft^2 ÷9 ft^2/yd^2 = 20.22 yd^2. Since Laurie cannot purchase part of a square yard, she has to round up. She must purchase 21 yd^2 to have enough to carpet the room.

367. **b.** One teaspoon equals 5 milliliters. Therefore, $\frac{3}{4}$ tsp × 5 mL/tsp = 3.75 mL.

368. **d.** One kiloliter equals 1,000 liters: 3.9 kL × 1,000 L/kL = 3,900 L. Each liter equals 1,000 milliliters: 3,900 L × 1,000 mL/L = 3,900,000 mL.

369. **c.** Each mile is about 1,609.34 meters. Divide to find out how many miles are equivalent to 1,500 meters: 1,500 m ÷ 1,609.34 m/mi ≈ 0.9321 mi.

370. **a.** There are 1,000 mm^3 in one cm^3. Divide to determine how many cm^3 are in 58.24 mm^3; 58.24 mm^3 ÷ 1,000 mm^3/cm^3 = 0.05824 cm^3.

371. **b.** Since one foot equals about 0.3048 meter, 5.5 feet equals: 5.5 ft × 0.3048 m/ft = 1.6764 m. Answer **a** is much too large. Now convert 1.6764 m into the remaining units given in the answer choices; 1.6764 m × 100 cm/m = 167.64 cm which is very close to answer **b.** 1.6764 m ÷ 1000 m/km = .0016764 km which eliminates answers **c** and **e**; 1.6764 m × 1,000 mm/m = 1,676.4 mm which is much higher than answer **d.** The answer is **b**: 170 cm.

372. **e.** One meter is approximately 3.281 feet. Therefore, 62.4 m × 3.281 ft/m ≈ 204.73ft ≈ 205 ft.

373. **c.** There are 10 millimeters in one centimeter: 84 mm ÷ 10 mm/cm = 8.4 cm, not 840 cm.

374. **d.** There are 1,000,000 square meters in one square kilometer. Divide: 350 m² ÷ 1,000,000 m²/km² = 0.00035 km².

375. **b.** Each liter equals 0.264 gallon. Multiply: 5 L × 0.264 gal/L = 1.32 gal.

376. **a.** One acre equals 4,840 square yards. Multiply: 0.16 acre × 4,840 yd²/acres = 774.4 yd².

377. **e.** Convert all units to meters first: 4.236 km × 1,000 m/km = 4,236 m; 654 cm ÷ 100 cm/m = 6.54 m. Now add: 4,236 m + 23 m + 6.54 m = 4,265.54 m.

378. **b.** One acre equals 0.004047 square kilometers, so divide: 5 km² ÷ 0.004047 km²/acres ≈ 1,235.48 acres.

379. **c.** One cubic foot equals 0.0283 cubic meter, so divide: 1 m³ ÷ 0.0283 m³/ft³ ≈ 35.3357 ft³ which rounds to 35.34 ft³.

380. **c.** One gallon is about 0.1337 cubic foot. Multiply: 14,360 gal × 0.1337 ft³/gal = 1,919.932 ft³ which is 1,920 ft³ rounded to the nearest cubic foot.

381. **e.** One yard equals about 0.9144 meter. Divide: 4.5 m ÷ 0.9144 m/yd ≈ 4.92 yd.

382. **a.** One square yard equals about 0.8361 square meter. Divide: 12.59 m² ÷ 0.8361 m²/yd² ≈ 15.06 yd².

383. **d.** One cubic yard is equivalent to about 0.7646 cubic meters. Multiply: 3 yd³ × 0.7646 m³/yd³ ≈ 2.2938 m³.

384. **c.** 2.957353 centiliters is equal to one fluid ounce. Divide: 15 cL ÷ 2.957353 cL/fl oz ≈ 5.1 fl oz.

385. **c.** Eight 8-oz glasses account for 64 ounces of fluid: 8 × 8 = 64. There are 128 ounces in a gallon. Divide: 64 oz ÷ 128 oz/gal = .5 gal.

386. **b.** There are 12 inches in one foot. Divide to find the number of candies in a foot: 12 in ÷ .5 in/m&m® = 24 m&m®. Every foot requires 24 candies and there are 5,280 feet in a mile: 5,280 ft/mi × 24 m&m®/ft = 126,720 m&m®/mi.

387. **e.** One square mile is equal to 640 acres. Multiply: 5,544 mi^2 × 640 acres/mi^2 = 3,548,160 acres.

388. **b.** There are 4 quarts in a gallon. Divide: 19 qts ÷ 4 qts/gal = 4.75 gal.

389. **d.** One meter equals 100 centimeters. Multiply: 16.9 m × 100 cm/m = 1,690 cm.

390. **a.** Since 2 pints equals 1 quart, 1 pint is equal to half a quart, or .5 quart. Therefore, 1 quart + 1 pint = 1 quart + .5 quart = 1.5 quarts. Each quart equals 0.946 liter, so multiply: 1.5 qts × 0.946 L/qt = 1.419 L.

391. **e.** One cubic inch is equivalent to 16.3871 cubic centimeters. Divide: 35 cm^3 ÷ 16.3871 cm^3/in^3 ≈ 2.14 in^3.

392. **b.** One square foot equals 0.0929 square meter. Multiply: 221 ft^2 × 0.0929 m^2/ft^2 ≈ 20.53 m^2.

393. **e.** There are 1,000 milliliters in one liter. Divide: 345.972 mL ÷ 1,000 mL/L = 0.345972 L.

394. **b.** One cubic yard is equal to 27 cubic feet. Each cubic foot is equal to 1,728 cubic inches. Multiply: 27 ft^3 × 1,728 in^3/ft^3 = 46,656 in^3.

395. **e.** One cup equals 8 fluid ounces; $2\frac{3}{4}$ cups = 2.75 cups. Multiply: 2.75 c × 8 fl oz/c = 22 fl oz.

396. **a.** One inch equals 2.54 centimeters. Multiply: 40 in × 2.54 cm/in = 101.6 cm.

397. **b.** One square centimeter equals 100 square millimeters. Multiply: 300 cm^2 × 100 mm^2/cm^2 = 30,000 mm^2.

398. **a.** Since one yard equals 0.9144 meter, multiply: 34 yd × 0.9144 m/yd ≈ 31.09 m.

399. **c.** One liter equals 0.264 gallon. Multiply: 2 L × 0.264 gal/L = 0.528 gal.

400. **d.** One square foot is equal to 144 square inches. Multiply: 13 ft² × 144 in²/ft² = 1,872 in².

401. **d.** One meter is equal to 100 centimeters. Multiply: 4.3 m × 100 cm/m = 430 cm.

402. **e.** To answer the question, subtract: 12.4 yds – 8.34 m. Since the answer must be given in meters, and the subtraction cannot be done unless the units on the two quantities match, convert 12.4 yards into meters. Since one yard = 0.9144 meter, multiply: 12.4 yds × 0.9144 m/yd ≈ 11.34 m. Now subtract: 11.34 m – 8.34 m = 3 m.

403. **b.** One square inch equals 6.4516 square centimeters. Multiply: 72 in² × 6.4516 cm²/in² ≈ 464.5 cm².

404. **d.** One cubic inch equals 16,387.064 cubic millimeters. Multiply: 11.8 in³ × 16,387.064 mm³/in³ ≈ 193,367.36 mm³.

405. **d.** One square meter is equal to 10,000 square centimeters. Divide: 29,728 cm² ÷ 10,000 cm²/m² = 2.9728 m². Fred should purchase 3 square meters in order to have enough felt.

406. **b.** One cubic meter equals 1,000,000 cubic centimeters. Multiply: 3.25 m³ × 1,000,000 cm³/m³ = 3,250,000 cm³.

407. **e.** One cup equals 8 fluid ounces. Multiply: 9 c × 8 fl oz/c = 72 fl oz.

408. **a.** One cubic foot equals 1,728 cubic inches. Divide: 226 in³ ÷ 1,728 in³/ft³ ≈ 0.13 ft³.

409. **c.** There are 1000 liters in one kiloliter. Divide: 1.5 L ÷ 1,000 L/kL = 0.0015 kL.

410. **c.** One quart equals 2 pints. Divide: 13 pts ÷ 2 pts/qt = 6.5 qts.

411. **b.** Ten millimeters equals 1 centimeter. Multiply: 13 cm × 10 mm/cm = 130 mm.

412. **d.** Convert 6,540 milliliters into liters, kiloliters, and ounces to find a match. There are 1,000 milliliters in one liter. Divide: 6,540 mL ÷ 1,000 mL/L = 6.54 L, which is answer choice **d**. The answer cannot be **a** since 6.54 L ≠ 0.00654 L. To convert 6,540 milliliters into kiloliters, divide: 6,540 mL ÷ 1,000,000 mL/kL = 0.00654 kL, which is smaller than answer choice **b**. To convert 6,540 milliliters into fluid ounces, divide: 6540 mL ÷ 29.57353 mL/fl oz = 221.14 fl oz which is not equal to answer choices **c** or **e**.

413. **a.** One mile equals 1.6093 kilometers so multiply: 4.21 mi × 1.6093 km/mi ≈ 6.775 km.

414. **e.** One fluid ounce equals 29.57 milliliters. Multiply: 10.5 fl oz × 29.57 mL/fl oz ≈ 310.49 mL.

415. **c.** One gallon equals 231 cubic inches. Divide: 450 in^3 ÷ 231 in^3/gal ≈ 1.948 gal.

416. **b.** One kilometer equals 1,000 meters. Divide: 4,876.2 m ÷ 1,000 m/km = 4.8762 km.

417. **a.** One square kilometer equals 1,000,000 square meters. Multiply: 4.45 km^2 × 1,000,000 m^2/km^2 = 4,450,000 m^2.

418. **a.** One liter equals 1,000 milliliters. Divide: 700 mL ÷ 1,000 mL/L = 0.7 L.

419. **d.** One inch equals 25.4 millimeters. Divide: 42 mm ÷ 25.4 mm/in ≈ 1.65 in.

420. **e.** One fluid ounce equals 0.00002957 cubic meters. Divide: 3 m^3 ÷ 0.00002957 m^3/fl oz ≈ 101,454 fl oz.

421. **e.** There are two ways to do this problem. One way is to find the area first, in square inches, then convert that into square feet: The area of the table is 53 in × 40 in = 2,120 in^2. Since one square foot equals 144 square inches, divide: 2,120 in^2 ÷ 144 in^2/ft^2 ≈ 14.72 ft^2. The second method involves converting the dimensions into feet first, then multiplying to get square feet. Since there are 12 inches in one foot, divide each dimension: 53 in ÷ 12 in/ft = 4.4167 ft. 40 in ÷ 12 in/ft = 3.3333 ft. Multiply: 4.4167 ft × 3.3333 ft = 14.72 ft^2.

422. **c.** One cubic centimeter equals 1,000 cubic millimeters. Multiply: 24 cm^3 × 1,000 mm^3/cm^3 = 24,000 mm^3.

423. **d.** One liter equals 1,000 milliliters. Multiply: 7.98 L × 1,000 mL/L = 7,980 mL.

424. **a.** One cubic yard equals 27 cubic feet. Divide: 765 ft^3 ÷ 27 ft^3/yd^3 ≈ 28.33 yd^3.

425. **e.** One gallon equals 4 quarts. Multiply: 231 gal × 4 qts/gal = 924 qts.

426. **b.** There are several pieces of information that can be used to answer this question. Here is one approach: since one kilometer equals 1,000 meters, multiply to find the number of meters in 2.8 kilometers; 2.8 km × 1,000 m/km = 2,800 m.

427. **c.** One kiloliter equals 1,000 liters. Multiply to find the number of liters: 12 kL × 1,000 L/kL = 12,000 L. Since each liter costs $0.05, multiply to find the cost: 12,000 L × $0.05/L = $600.

428. **c.** One gallon equals about 3.785 liters. Multiply: 2.5 gal × 3.785 L/gal = 9.4625 L.

429. **a.** One square mile equals about 2.59 square kilometers. Multiply: 3 mi^2 × 2.59 km^2/mi^2 = 7.77 km^2.

430. **d.** One quart equals 0.946 liter. Multiply: 34 qts × 0.946 L/qt = 32.164 L.

431. **d.** One cubic inch equals 16,387.064 cubic millimeters. Multiply: 24 in^3 × 16,387.064 mm^3/in^3 = 393,289.536 mm^3.

432. **b.** Find a common unit and convert all measurements to this unit. If meters are chosen, for example, 540 m will remain the same. 7,500 cm = 75 m since there are 100 centimeters in each meter: 7,500 cm ÷ 100 cm/m = 75 m. 0.5 km = 500 m since there are 1,000 meters in every kilometer: 0.5 km × 1,000 m/km = 500 m. 4,000 mm = 4 m since there are 1,000 millimeters in every meter: 4,000 mm ÷ 1,000 mm/m = 4 m. Now compare: 540 m, 75 m, 500 m, 4 m. 540 m is the largest amount, followed by 500 m, then 75 m, then 4 m. Answer choice **b** indicates this order.

433. **e.** One square centimeter equals 100 square millimeters. Divide: 4,500 mm^2 ÷ 100 mm^2/cm^2 = 45 cm^2.

434. **a.** Find a common unit and convert all measurements to this unit. If liters are chosen, for example, 20 L will remain the same. 0.2 kL = 200 L since there are 1,000 liters in every kiloliter: 0.2 kL × 1,000 L/kL = 200 L. 200,000 mL = 200 L since there are 1,000 milliliters in every liter: 200,000 mL ÷ 1,000 mL/L = 200 L. Now compare: 20 L, 200 L, and 200 L. 20 L is the smallest amount.

435. **c.** One pint equals 2 cups. Divide: 75 c ÷ 2 c/pt = 37.5 pt.

436. **a.** One mile equals 1,609.34 meters. Divide: 2,500 m ÷ 1,609.347 m/mi ≈ 1.5534 mi. Round to the nearest tenth to get 1.6 mi.

437. **b.** One liter equals 1.057 quarts. Multiply: 2.2 L × 1.057 qts/L = 2.3254 qts. Each quart equals 2 pints so multiply: 2.3254 qts × 2 pts/qt = 4.6508 pts. Each pint equals 2 cups so multiply one last time to convert into cups: 4.6508 pts × 2 c/pt = 9.3016 c.

438. **a.** One cubic meter equals 1,000,000 cubic centimeters. Divide: 9,760,000 cm^3 ÷ 1,000,000 cm^3/m^3 = 9.76 m^3.

439. **d.** One liter equals 1,000 milliliters. Divide: 1,500 mL ÷ 1,000 mL/L = 1.5 L.

440. **e.** One foot equals about 0.3048 meter. Multiply: 4,321 ft × 0.3048 m/ft ≈ 1,317 m.

441. **c.** One cubic meter equals 1,000,000 cubic centimeters. Multiply: 4,000,000 m^3 × 1,000,000 cm^3/m^3 = 4,000,000,000,000 cm^3.

10

Conversion—
Metric and Standard Units:
Temperature and Weight

This final chapter covers the conversions for weight and temperature. Much of the world measures temperature in degrees Celsius. The Celsius unit of temperature is based on the freezing point and boiling point of water. Water freezes at 0°C and boils at 100°C, at sea level. In the United States, temperature is measured in degrees Fahrenheit. Refer to the formula sheet for the formula used to convert from one type of unit to the other. These conversions will be practiced and explained in the questions and answer explanations.

Weight is measured in metric units. The basic unit is the gram, and the other units in this system (centigram, milligram, and kilogram), are based on the powers of ten. This was dealt with in Chapter 9, with the basic units of meter and liter. Converting within the metric system is relatively easy and is outlined in the formula sheet. Standard units of weight are not so straightforward. Weight in this system is measured in ounces, pounds, and tons. These units are defined in the formula sheet, as well as conversion formulas to convert between the metric system and the standard system.

Work through the 60 problems in this chapter using the formula sheet as your guide. For all problems, the answer explanations will outline the methods to use to perform these conversions.

Formulas

Weight conversions between standard units

1 pound (lb) = 16 ounces (oz)
1 ton (t) = 2,000 pounds

Weight conversions between metric units

1 centigram (cg) = 10 milligrams (mg)
1 gram (g) = 100 centigrams = 1,000 milligrams
1 kilogram (kg) = 1,000 grams

Weight conversions between standard and metric units

1 gram ≈ 0.035 ounce
1 pound ≈ 0.454 kilogram
1 pound ≈ 454 grams
1 kilogram ≈ 2.205 pounds
1 ton ≈ 908 kilograms

Temperature conversions from Celsius to Fahrenheit

To convert from degrees Celsius (°C) to degrees Fahrenheit (°F), use the formula:

$$F = \frac{9}{5}C + 32$$

Substitute the given number of Celsius degrees in the formula for C.
Multiply by $\frac{9}{5}$ and then add 32.

Example:
Convert 40°C into Fahrenheit.
$F = \frac{9}{5}(40) + 32$
$F = \frac{360}{5} + 32$
$F = 72 + 32$
$F = 104°$
Therefore, 40°C = 104°F

Temperature conversions from Fahrenheit to Celsius

To convert from degrees Fahrenheit (°F) to degrees Celsius (°C), use the formula:

$$C = \tfrac{5}{9}(°F - 32)$$

Substitute the given number of Fahrenheit degrees in the formula for F. Subtract 32, then multiply by $\tfrac{5}{9}$.

Example:
Convert 50°F into Celsius.
$C = \tfrac{5}{9}(50 - 32)$
$C = \tfrac{5}{9}(18)$
$C = \tfrac{90}{9}$
$C = 10°$
Therefore, 50°F = 10°C

442. 6 lb = _____ oz
   a. 96
   b. 69
   c. 2.667
   d. 0.375
   e. 22

443. It is a commonly known fact that 0°C = 32°F, the freezing point of water. However, the conversion from 0°F to degrees is not as commonly known. Find the equivalent Celsius temperature for zero degrees Fahrenheit.
   a. 17.8°C
   b. −17.8°C
   c. −32°C
   d. 32°C
   e. −57.6°C

444. Tommy drives a half-ton pickup truck. If *half ton* refers to the amount of weight the truck can carry in its bed, how many pounds can the truck carry?
   a. 8 lb
   b. 100 lb
   c. 2,000 lb
   d. 1,000 lb
   e. 4,000 lb

445. 48 kg = _____ g
   a. 0.0048
   b. 0.48
   c. 4.8
   d. 4,800
   e. 48,000

446. If the temperature is 10 degrees below zero Fahrenheit, what is the temperature in degrees Celsius?
   a. 12.2°C
   b. −37.6°C
   c. 26.4°C
   d. −23.3°C
   e. 14°C

447. 3 kg ≈ _____ lbs
   a. 1.4
   b. 6.6
   c. 5.2
   d. 3.5
   e. 2.5

448. 845.9 cg = _____ kg
   a. 0.008459
   b. 0.8450
   c. 8.459
   d. 84.59
   e. 84,590

449. 104°F = _____ °C
   a. 75.6
   b. 25.8
   c. 89.8
   d. 219.2
   e. 40

450. 2.5 lbs ≈ _____ g
   a. 1.816
   b. 1.135
   c. 1135
   d. 1816
   e. 5.51

451. With the wind chill, the weatherman reported a temperature of 22 degrees below zero. Write this temperature in degrees Celsius.
   a. 5.6°C
   b. −44.2°C
   c. 19.8°C
   d. −30°C
   e. −7.6°C

452. A recipe calls for 3 ounces of cheese per serving. If Rebecca is making 12 servings, how many pounds of cheese will she need?
    a. 3.6 lbs
    b. 36 lbs
    c. 2.25 lbs
    d. 22.5 lbs
    e. 4.5 lbs

453. Graham cracker sticks come in a 13-ounce package. What is the approximate weight of the package in kilograms?
    a. 0.371 kilograms
    b. 371.4 kilograms
    c. 0.455 kilograms
    d. 455 kilograms
    e. 286.3 kilograms

454. 432 ounces = _____ tons
    a. 27
    b. 2.7
    c. 0.135
    d. 0.0135
    e. 0.216

455. $-12°C =$ _____ °F
    a. −79.2
    b. 10.4
    c. 36
    d. −53.6
    e. −24.4

456. "Take 500 milligrams of ibuprofen if your headache acts up," the doctor ordered. How many grams of ibuprofen is this?
    a. 0.05 grams
    b. 5,000 grams
    c. 50 grams
    d. 5 grams
    e. 0.5 grams

457. 4.2 t ≈ _____ kg

    a. 3,813.6

    b. 38.136

    c. 216.2

    d. 21.62

    e. 2,162

458. A Galapagos tortoise can weigh up to 300 pounds. Convert this weight into tons.

    a. 60 tons

    b. 6 tons

    c. 6.67 tons

    d. 1.5 tons

    e. 0.15 tons

459. 220°F is above the boiling point of water. What is this temperature in degrees Celsius?

    a. 104.4°C

    b. 428°C

    c. 90.2°C

    d. 140°C

    e. 154.2°C

460. 3.15 t = _____ oz

    a. 100,800

    b. 10,080

    c. 50

    d. 6,300

    e. 393.75

461. A box of Jello weighs about 85 grams. Convert this weight to kilograms.

    a. 85,000 kilograms

    b. 850 kilograms

    c. 0.85 kilograms

    d. 0.085 kilograms

    e. 0.0085 kilograms

462. $1.4 \, \text{kg} \approx$ _____ oz
  a. 0.049
  b. 0.1929
  c. 22
  d. 11
  e. 49

463. Susan bought a 25-gram container of ground cloves to use in cider. How many milligrams of cloves did she buy?
  a. 250 milligrams
  b. 2,500 milligrams
  c. 25,000 milligrams
  d. 2.5 milligrams
  e. 0.025 milligrams

464. $58 \, \text{g} \approx$ _____ oz
  a. 2.03
  b. 1,657
  c. 26.332
  d. 127.75
  e. 48.13

465. Josephine has 70 ounces of meat to use in her annual batch of spaghetti sauce. How many pounds of meat does Josephine have?
  a. 4.375 pounds
  b. 1,120 pounds
  c. 8.75 pounds
  d. 3.5 pounds
  e. 5.13 pounds

466. $-85°\text{F} =$ _____ °C
  a. −121
  b. −15.2
  c. −29.4
  d. −65
  e. −79.2

467. 63 mg = _____ cg

    a. 630

    b. 63

    c. 6.3

    d. 0.63

    e. 0.063

468. 2,800 g + 390 cg + 2 kg ≈ _____ lbs

    a. 3,192

    b. 7.03

    c. 10.58

    d. 19.16

    e. 3.9

469. Mary's teeth were chattering. It was −5°F outside and she was cold to the bone. Convert this temperature to degrees Celsius.

    a. 23°C

    b. −34.8°C

    c. 29.2°C

    d. −20.6°C

    e. 15°C

470. Fill in the missing value in the table below so that the values are equivalent:

Kilograms	Milligrams
	1,254,600

    a. 1,254,600 kilograms

    b. 1.2546 kilograms

    c. 1,254,600,000 kilograms

    d. 1,254,600,000,000 kilograms

    e. 12.546 kilograms

471. Kelly's dog weighs 52 pounds. Convert the dog's weight to kilograms.
   a. 23,608
   b. 8.73
   c. 236.1
   d. 114.5
   e. 23.608

472. 20°C = _____ °F
   a. −6.67
   b. 68
   c. −21.6
   d. 93.6
   e. 4

473. The oven thermometer read 65°C during preheating. Convert this temperature to Fahrenheit.
   a. 174.6°F
   b. 18.33°F
   c. 85°F
   d. 149°F
   e. 59.4°F

474. Jodee feeds her daughter 4 ounces of baby food at every meal. About how many grams of food does Jodee's daughter get at each meal?
   a. 0.14 grams
   b. 4.035 grams
   c. 3.965 grams
   d. 114.3 grams
   e. 120.2 grams

475. 90°C = _____ °F
   a. 194
   b. 130
   c. 32.2
   d. 104.4
   e. 219.6

476. Convert –10°C into degrees Fahrenheit.
a. 14°F
b. –50°F
c. –23.3°F
d. 39.6°F
e. –75.6°F

477. –35°C = _____ °F
a. –95
b. –120.6
c. –37.2
d. –95
e. –31

478. The thermostat was set at 17°C for the spring. Convert this temperature into degrees Fahrenheit.
a. –27°F
b. 88.2°F
c. 62.6°F
d. –8.3°F
e. –1.4°F

479. 2°C = _____ °F
a. –16.7
b. –28.4
c. –54
d. 61.2
e. 35.6

480. Julie was hoping the temperature would reach 26°C tomorrow. What is this temperature in degrees Fahrenheit?
a. 104.4°F
b. –10.8°F
c. 14.8°F
d. 78.8°F
e. –3.3°F

481. 34°C = _____ °F
   a. 93.2
   b. 3.6
   c. 1.1
   d. 29.2
   e. 118.8

482. Temperatures in some parts of the world can be 57°C. Convert this temperature into degrees Fahrenheit.
   a. 160.2°F
   b. 45°F
   c. 134.6°F
   d. 13.9°F
   e. 70.6°F

483. Convert –81°C into degrees Fahrenheit.
   a. –203.4°F
   b. –113.8°F
   c. –177.8°F
   d. –62.8°F
   e. –88.2°F

484. 12.71 kilograms = _____ milligrams
   a. 0.01271
   b. 0.00001271
   c. 12,710
   d. 1,271,000
   e. 12,710,000

485. –63°C = _____ °F
   a. –52.8
   b. –145.4
   c. –171
   d. –81.4
   e. –55.8

486. The thermometer reading of –40°C is interesting. Convert this temperature into degrees Fahrenheit to determine why.
   a. 140°F
   b. –40°F
   c. 40°F
   d. 0°F
   e. –104°F

487. A doctor prescribed 2 centigrams of medicine to his patient. How many milligrams of medicine did the doctor prescribe?
   a. 2,000 milligrams
   b. 200 milligrams
   c. 20 milligrams
   d. 2 milligrams
   e. 0.2 milligrams

488. –7°C = _____ °F
   a. –44.6
   b. –21.7
   c. 19.4
   d. –70.2
   e. 45

489. 0.24 kg = _____ cg
   a. 0.00024
   b. 240
   c. 24
   d. 24,000
   e. 0.0000024

490. The freezing point of water is 32°F. Convert this into degrees Celsius.
   a. 17.8°C
   b. –32°C
   c. 0°C
   d. 32°C
   e. –57.6°C

491. The temperature dropped to 5°F during the evening. What is this temperature in degrees Celsius?
   a. −15°C
   b. −29.2°C
   c. 34.8°C
   d. 20.6°C
   e. 41°C

492. 183°F = _____ °C
   a. 69.7
   b. 83.9
   c. 133.7
   d. 361.4
   e. 119.4

493. 15°F = _____ °C
   a. −23.7
   b. −9.4
   c. 59
   d. 26.1
   e. 40.3

494. Michael weighed 12 pounds 9 ounces at his last checkup. Convert Michael's weight into ounces.
   a. 148 oz
   b. 21 oz
   c. 9.75 oz
   d. 156 oz
   e. 201 oz

495. The air temperature was a humid 97°F on the beach. Convert this temperature into degrees Celsius.
   a. 206.6°C
   b. 36.1°C
   c. 71.7°C
   d. 21.9°C
   e. 85.9°C

496. 29°F = _____ °C

    a. −15.9

    b. 48.1

    c. −1.7

    d. 84.2

    e. 33.9

497. _____ cg = 564.9 g

    a. 5.649

    b. 56,490

    c. 5,649

    d. 56.49

    e. 0.5649

498. −19°F = _____ °C

    a. 7.2

    b. −42.6

    c. 21.4

    d. −2.2

    e. −28.3

499. The empty container weighs 512 centigrams; 42,370 milligrams of sand are added to the cup. What is the total weight of the sand and the container in grams?

    a. 554.37 grams

    b. 42,882 grams

    c. 47.49 grams

    d. 93.57 grams

    e. 9.357 grams

500. −21°F = _____ °C

    a. −5.8

    b. 6.1

    c. −43.7

    d. 20.3

    e. −29.4

501. An African Elephant can weigh as much as 5,448 kilograms. Convert the elephant's weight into tons.

    a. 2.7 tons

    b. 6.0 tons

    c. 3.5 tons

    d. 2.2 tons

    e. 1.8 tons

Answers

442. **a.** Every pound equals 16 ounces, so multiply: 6 lb × 16 oz/lb = 96 oz.

443. **b.** To convert 0°F into Celsius, substitute 0 for F in the equation:
$C = \frac{5}{9}(F - 32)$
$C = \frac{5}{9}(0 - 32)$
$C = \frac{5}{9}(-32)$
$C = -\frac{160}{9}$
$C \approx -17.8°$
Therefore, 0°F ≈ −17.8°C.

444. **d.** One ton equals 2,000 pounds. Therefore, half of a ton is 2,000 ÷ 2 = 1,000 pounds.

445. **e.** Every kilogram equals 1,000 grams, so multiply: 48 kg × 1,000 g/kg = 48,000 g.

446. **d.** To convert −10°F into Celsius, substitute −10 for F in the equation:
$C = \frac{5}{9}(F - 32)$
$C = \frac{5}{9}(-10 - 32)$
$C = \frac{5}{9}(-42)$
$C = \frac{-210}{9}$
$C \approx -23.3°$
Therefore, −10°F ≈ −23.3°C.

447. **b.** Each kilogram equals about 2.205 pounds. Since there are 3 kilograms, multiply: 3 kg × 2.205 lbs/kg = 6.615 lbs ≈ 6.6 lbs when rounded to the nearest tenth.

448. **a.** Every 100,000 centigrams make a kilogram so divide: 845.9 cg ÷ 100,000 cg/kg = 0.008459 kg.

449. **e.** To convert 104°F into Celsius, substitute 104 for F in the equation:
$C = \frac{5}{9}(F - 32)$
$C = \frac{5}{9}(104 - 32)$
$C = \frac{5}{9}(72)$
$C = \frac{360}{9}$
$C = 40°$
Therefore, 104°F = 40°C.

450. **c.** Every pound equals 454 grams. Therefore, to convert 2.5 pounds into grams, multiply: 2.5 lbs × 454 g/lb = 1,135 g.

451. **d.** To convert –22°F into Celsius, substitute –22 for F in the equation:

$$C = \tfrac{5}{9}(F - 32)$$
$$C = \tfrac{5}{9}(-22 - 32)$$
$$C = \tfrac{5}{9}(-54)$$
$$C = \tfrac{-270}{9}$$
$$C = -30°$$

Therefore, –22°F = –30°C.

452. **c.** Since each serving calls for 3 ounces of cheese, and Rebecca is making 12 servings, she needs: 3 oz × 12 = 36 oz of cheese. Every 16 ounces is equal to one pound, so divide to convert ounces into pounds: 36 oz ÷ 16 oz/lb = 2.25 lb.

453. **a.** Every 0.035 ounce is equal to 1 gram, so first figure out how many grams are in 13 ounces by dividing: 13 oz ÷ 0.035 oz/g ≈ 371.4 g. Since 1,000 g equals 1 kg, divide to determine the number of kilograms: 371.4 g ÷ 1,000g/kg ≈ 0.371 kg.

454. **d.** 16 ounces equals 1 pound, so divide: 432 oz ÷ 16 oz/lb = 27 lb. It takes 2,000 pounds to make 1 ton, so divide to discover what fraction of a ton 27 pounds is: 27 lb ÷ 2,000 lb/t = 0.0135 t.

455. **b.** To convert –12°C into Fahrenheit, substitute –12 for C in the equation:

$$F = \tfrac{9}{5}C + 32$$
$$F = \tfrac{9}{5}(-12) + 32$$
$$F = \tfrac{-108}{5} + 32$$
$$F = -21.6 + 32$$
$$F = 10.4°$$

Therefore, –12°C = 10.4°F.

456. **e.** 1,000 milligrams equals 1 gram, so divide: 500 mg ÷ 1,000 mg/g = 0.5 grams.

457. **a.** Every ton equals 908 kilograms. Multiply: 4.2 t × 908 kg/t = 3,813.6 kg.

458. **e.** 2,000 pounds equals 1 ton, so divide: 300 lb ÷ 2,000 lb/t = 0.15 t.

459. **a.** To convert 220°F into Celsius, substitute 220 for F in the equation:
$C = \frac{5}{9}(F - 32)$
$C = \frac{5}{9}(220 - 32)$
$C = \frac{5}{9}(188)$
$C = \frac{940}{9}$
$C \approx 104.4°$
Therefore, 220°F ≈ 104.4°C.

460. **a.** Every ton equals 2,000 pounds, so multiply to find 3.15 pounds: 3.15 t × 2,000 lbs/t = 6,300 lbs. Each pound equals 16 ounces, so multiply again: 6,300 lbs × 16 oz/lb = 100,800 oz.

461. **d.** 1,000 grams equals one kilogram, so divide to convert 85 grams: 85 g ÷ 1,000g/kg = 0.085 kg.

462. **e.** Each kilogram is equivalent to 2.205 pounds, so multiply: 1.4 kg × 2.205 lb\kg = 3.087 lbs. Each pound equals 16 ounces, so multiply again: 3.087 lbs × 16 oz/lb = 49.392 oz. Alternately, you can first convert kilograms to grams, and then to pounds. Each kilogram is equal to 1,000 grams, so kilograms is 1.4 × 1,000 = 1,400 grams. Each gram is 0.035 ounces, so multiply; 1,400 × .0.035 = 49 oz.

463. **c.** Each gram equals 1,000 milligrams. Multiply: 25 g × 1,000 mg/g = 25,000 mg.

464. **a.** Each gram equals about 0.035 ounce. Multiply: 58 g × 0.035 oz/g = 2.03 oz.

465. **a.** Every 16 ounces make one pound, so divide: 70 oz ÷ 16 oz/lb = 4.375 lb.

466. **d.** To convert –85°F into Celsius, substitute –85 for F in the equation:
$C = \frac{5}{9}(F - 32)$
$C = \frac{5}{9}(-85 - 32)$
$C = \frac{5}{9}(-117)$
$C = \frac{-585}{9}$
$C = -65°$
Therefore, –85°F = –65°C.

467. **c.** Ten milligrams equals one centigram, so divide: 63 mg ÷ 10 mg/cg = 6.3 cg.

468. **c.** Convert all units to a single unit first, grams for example: 2,800 g remains the same; 390 cg ÷ 100 cg/g = 3.9 g; 2 kg × 1,000 g/kg = 2,000 g. Now add all the converted values: 2,800 g + 3.9 g + 2,000 g = 4,803.9 g. Every 454 grams equals 1 pound, so divide: 4,803.9 g ÷ 454 g/lb ≈ 10.58 lb.

469. **d.** To convert −4°F into Celsius, substitute −4 for F in the equation:
$C = \frac{5}{9}(F - 32)$
$C = \frac{5}{9}(-5 - 32)$
$C = \frac{5}{9}(-37)$
$C = \frac{-185}{9}$
$C \approx -20.6°$
Therefore, −4°F ≈ −20.6°C.

470. **b.** Every 1,000 milligrams equals 1 gram, so divide to determine the number of grams: 1,254,600 mg ÷ 1,000 mg/g = 1,254.6 g. Now, every 1,000 grams equals 1 kilogram, so divide again: 1,254.6 g ÷ 1,000 g/kg = 1.2546 kg.

471. **e.** Each pound equals 0.454 kilograms, so multiply to find the number of kilograms in 52 pounds: 52 lbs × 0.454 kg/lb = 23.608 kg.

472. **b.** To convert 20°C into Fahrenheit, substitute 20 for C in the equation:
$F = \frac{9}{5}C + 32$
$F = \frac{9}{5}(20) + 32$
$F = \frac{180}{5} + 32$
$F = 36 + 32$
$F = 68°$
Therefore, 20°C = 68°F.

473. **d.** To convert 65°C into Fahrenheit, substitute 65 for C in the equation:
$F = \frac{9}{5}C + 32$
$F = \frac{9}{5}(65) + 32$
$F = \frac{585}{5} + 32$
$F = 117 + 32$
$F = 149°$
Therefore, 65°C = 149°F.

474. **d.** Every 0.035 ounce is equal to 1 gram, so divide 4 ounces by 0.035 ounces to determine how many grams are in 4 ounces: 4 oz ÷ 0.035 oz/g = 114.2857 g ≈ 114.3 g.

475. **a.** To convert 90°C into Fahrenheit, substitute 90 for C in the equation:
$F = \frac{9}{5}C + 32$
$F = \frac{9}{5}(90) + 32$
$F = \frac{810}{5} + 32$
$F = 162 + 32$
$F = 194°$
Therefore, 90°C = 194°F.

476. **a.** To convert −10°C into Fahrenheit, substitute −10 for C in the equation:
$F = \frac{9}{5}C + 32$
$F = \frac{9}{5}(-10) + 32$
$F = \frac{-90}{5} + 32$
$F = -18 + 32$
$F = 14°$
Therefore, −10°C = 14°F.

477. **e.** To convert −35°C into Fahrenheit, substitute −35 for C in the equation:
$F = \frac{9}{5}C + 32$
$F = \frac{9}{5}(-35) + 32$
$F = \frac{-315}{5} + 32$
$F = -63 + 32$
$F = -31°$
Therefore, −35°C = −31°F.

478. **c.** To convert 17°C into Fahrenheit, substitute 17 for C in the equation:
$F = \frac{9}{5}C + 32$
$F = \frac{9}{5}(17) + 32$
$F = \frac{153}{5} + 32$
$F = 30.6 + 32$
$F = 62.6°$
Therefore, 17°C = 62.6°F.

479. **e.** To convert 2°C into Fahrenheit, substitute 2 for C in the equation:
$F = \frac{9}{5}C + 32$
$F = \frac{9}{5}(2) + 32$
$F = \frac{18}{5} + 32$
$F = 3.6 + 32$
$F = 35.6°$
Therefore, 2°C = 35.6°F.

480. **d.** To convert 26°C into Fahrenheit, substitute 26 for C in the equation:
$$F = \tfrac{9}{5}C + 32$$
$$F = \tfrac{9}{5}(26) + 32$$
$$F = \tfrac{234}{5} + 32$$
$$F = 46.8 + 32$$
$$F = 78.8°$$
Therefore, 26°C = 78.8°F.

481. **a.** To convert 34°C into Fahrenheit, substitute 34 for C in the equation:
$$F = \tfrac{9}{5}C + 32$$
$$F = \tfrac{9}{5}(34) + 32$$
$$F = \tfrac{306}{5} + 32$$
$$F = 61.2 + 32$$
$$F = 93.2°$$
Therefore, 34°C = 93.2°F.

482. **c.** To convert 57°C into Fahrenheit, substitute 57 for C in the equation:
$$F = \tfrac{9}{5}C + 32$$
$$F = \tfrac{9}{5}(57) + 32$$
$$F = \tfrac{513}{5} + 32$$
$$F = 102.6 + 32$$
$$F = 134.6°$$
Therefore, 57°C = 134.6°F.

483. **b.** To convert −81°C into Fahrenheit, substitute −81 for C in the equation:
$$F = \tfrac{9}{5}C + 32$$
$$F = \tfrac{9}{5}(-81) + 32$$
$$F = \tfrac{-729}{5} + 32$$
$$F = -145.8 + 32$$
$$F = -113.8°$$
Therefore, −81°C = −113.8°484. e. One kilogram equals 1,000 grams, so multiply: 12.71 kg × 1,000 g/kg = 12,710 g. One gram equals 1,000 milligrams, so multiply again: 12,710 g × 1,000 mg/g = 12,710,000 mg.

485. **d.** To convert –63°C into Fahrenheit, substitute –63 for C in the equation:

$F = \frac{9}{5}C + 32$

$F = \frac{9}{5}(-63) + 32$

$F = \frac{-567}{5} + 32$

$F = -113.4 + 32$

$F = -81.4°$

Therefore, –63°C = –81.4°486. b. To convert –40°C into Fahrenheit, substitute –40 for C in the equation:

$F = \frac{9}{5}C + 32$

$F = \frac{9}{5}(-40) + 32$

$F = \frac{-360}{5} + 32$

$F = -72 + 32$

$F = -40°$

Therefore, –40°C = –40°F.

487. **c.** Each centigram equals 10 milligrams. Therefore, to find the number of milligrams in 2 centigrams, multiply: 2 cg $\times$ 10 mg/cg = 20 mg.

488. **c.** To convert –7°C into Fahrenheit, substitute –7 for C in the equation:

$F = \frac{9}{5}C + 32$

$F = \frac{9}{5}(-7) + 32$

$F = \frac{-63}{5} + 32$

$F = -12.6 + 32$

$F = 19.4°$

Therefore, –7°C = 19.4°F.

489. **d.** Each kilogram is 1,000 grams, so multiply to find the number of grams in 0.24 kg: 0.24 kg $\times$ 1,000 g/kg = 240 g. Each gram equals 100 centigrams so multiply again: 240 g $\times$ 100 cg/g = 24,000 cg.

490. **c.** To convert 32°F into Celsius, substitute 32 for F in the equation:

$C = \frac{5}{9}(F - 32)$

$C = \frac{5}{9}(32 - 32)$

$C = \frac{5}{9}(0)$

$C = \frac{0}{9}$

$C = 0°$

Therefore, 32°F = 0°C.

491. **a.** To convert 5°F into Celsius, substitute 5 for F in the equation:
$C = \frac{5}{9}(F - 32)$
$C = \frac{5}{9}(5 - 32)$
$C = \frac{5}{9}(-27)$
$C = \frac{-135}{9}$
$C = -15°$
Therefore, 5°F = −15°C.

492. **b.** To convert 183°F into Celsius, substitute 183 for F in the equation:
$C = \frac{5}{9}(F - 32)$
$C = \frac{5}{9}(183 - 32)$
$C = \frac{5}{9}(151)$
$C = \frac{755}{9}$
$C \approx 83.9°$
Therefore, 183°F ≈ 83.9°C.

493. **b.** To convert 15°F into Celsius, substitute 15 for F in the equation:
$C = \frac{5}{9}(F - 32)$
$C = \frac{5}{9}(15 - 32)$
$C = \frac{5}{9}(-17)$
$C = \frac{-85}{9}$
$C \approx -9.4°$
Therefore, 15°F ≈ −9.4°C.

494. **e.** Every pound equals 16 ounces, so 12 pounds would equal: 12 lbs × 16 oz/lb = 192 oz. Since Michael weighed 12 pounds 9 ounces, add the 9 ounces to the number of ounces in 12 pounds: 192 ounces + 9 ounces = 201 ounces.

495. **b.** To convert 97°F into Celsius, substitute 97 for F in the equation:
$C = \frac{5}{9}(F - 32)$
$C = \frac{5}{9}(97 - 32)$
$C = \frac{5}{9}(65)$
$C = \frac{325}{9}$
$C \approx 36.1°$
Therefore, 97°F ≈ 36.1°C.

496. **c.** To convert 29°F into Celsius, substitute 29 for F in the equation:

$C = \frac{5}{9}(F - 32)$

$C = \frac{5}{9}(29 - 32)$

$C = \frac{5}{9}(-3)$

$C = -\frac{15}{9}$

$C \approx -1.7°$

Therefore, 29°F $\approx$ −1.7°C.

497. **b.** Each gram is 100 centigrams, so multiply: 564.9 g × 100 cg/g = 56,490 cg.

498. **e.** To convert −19°F into Celsius, substitute −19 for F in the equation:

$C = \frac{5}{9}(F - 32)$

$C = \frac{5}{9}(-19 - 32)$

$C = \frac{5}{9}(-51)$

$C = \frac{-255}{9}$

$C \approx -28.3°$

Therefore, −19°F $\approx$ −28.3°C.

499. **c.** 512 centigrams is equal to 5.12 grams since every 100 centigrams equals 1 gram: 512 cg ÷ 100 cg/g = 5.12 g. 42,370 milligrams equals 42.370 grams since every 1,000 milligrams equals 1 gram: 42,370 mg ÷ 1,000 mg/g = 42.37 g. Now add: 5.12 g + 42.37 g = 47.49 g.

500. **e.** To convert −21°F into Celsius, substitute −21 for F in the equation:

$C = \frac{5}{9}(F - 32)$

$C = \frac{5}{9}(-21 - 32)$

$C = \frac{5}{9}(-53)$

$C = -\frac{265}{9}$

$C \approx -29.4°$

Therefore, −21°F $\approx$ −29.4°C.

501. **b.** Every 908 kilograms is equivalent to 1 ton, so divide: 5,448 kg ÷ 908 kg/t = 6 t.